The Trees of Mount Rainier

by

Susan McDougall

Sound Books

Seattle, Washington

To those who venture into the forest and find beauty there.

McDougall, Susan.
The Trees of Mount Rainier/ by Susan McDougall.
2nd ed.
Seattle, Wash.: Sound Books, c2007.
1. Trees--Washington (State)--Rainier, Mount. 2. Botany--Washington (State)--Rainier, Mount. 3. Adams, Mount (Wash.).
I. McDougall, Susan.

ISBN 978-0-9776285-2-0

Second edition 2007

Sound Books

13008 37th Ave. NE
Seattle, WA 98125

http://mtadamsflora.clearwire.net

Contents

Contents

Preface

Mount Rainier from the south

In the heart of the Pacific Northwest a great volcano smolders beneath an icy cloak. Molded by glaciers and perched more than 14,000 feet above a lowland plain, on modern maps the mountain is named "Rainier" in honor of a man who never saw it. To the earliest inhabitants it was "Tahoma"; to residents within its sphere of influence, it is "the mountain."

Mount Rainier dominates the surrounding landscape. It is visible from bridges that cross great lakes in bustling cities; it pokes above urbanites' back yards, and creeps into view from high ridges a hundred miles away. Round a corner on a mountain trail, and there it is, taking the hiker a bit by surprise. Ominously close to densely populated lands, the mountain has inflicted its wrath on former residents, burying river valleys in giant swaths of mud, making the land fertile for future inhabitants. Within the mountain, volcanic fires have flared repeatedly, decapitating old summits and remaking the landscape. Climb to the top and seek shelter within steamy caves kept warm by internal heat. In forested valleys below, hot springs hint that the aging volcano still lives.

Time and the elements have taken their toll on the mountain. Carved by 25 named glaciers, and sluiced by rain and snow, with each passing day crumbled rock is carried seaward by giant rivers. Of much greater extent in the past, the glaciers have sculpted an eroded hulk. But even this dynamic environment fails to obscure the uniqueness of the mountain. It stands apart from the Cascades range to the east, and dwarfs its volcanic companions. Climb it, and the surrounding mountains appear as mere foothills. At the summit, the world drops away.

The wild and colorful landscape is beautiful enough to warrant national park status, where over a million people visit each year. The raw power and beauty of the mountain is at least in part defined by the forests that surround it. Providing a contrast to the icy cone, dark green trees line high ridges, pressing against the moving glaciers.

Unique to the Northwest, the forest communities of Mount Rainier encompass both stability and change. Dynamic forces that have impacted the forest members in the past continue to do so today.

This book is about the communities of trees that constitute those forests. Written for the layperson, the text keeps technical terms to a minimum, and although the book has an educational intent, of equal importance is the hope to enhance enjoyment of the mountain. Crossing the Park boundary, an awareness of the difference between the new landscape and the old is immediately apparent. Much of that impact resides in the immense and varied forests.

The text explains the dynamics of the interactions between the tree species and the physical and biological environment that influences their distribution. Examples of particular species and their forests are given to clarify and augment the discussion. The chapters are intended to stand alone in content, so that portions of the book may be selected, if desired.

Longmire forests

The first chapter, "The Forest Visited," introduces the tree species of Mount Rainier and briefly discusses the factors that influence forest composition.

Chapter Two, "The Volcano and its Forests," begins with an overview of the geology and glaciology of the mountain. As the highest volcano in the lower 48 states, Mount Rainier has been subjected to repeated episodes of volcanic and glacial activity; these forces have defined and shaped the topography, as well as the living communities of the region. The chapter concludes with a discussion of the role of those forces in the composition of the forests that encircle the mountain.

The third chapter, "The Trees," begins with an introduction to the two types of trees represented at Mount Rainier — the needle-bearing coniferous species, and the leaf-bearing species. Following this overview, a description of the several forest community types is discussed. A presentation of the evolutionary history of the represented trees provides more background on the two types of trees. Finally, the discussion of tree parts, such as roots and branches, is preceded by a section on soil, the organism upon which the trees so strongly depend.

The fourth chapter, "Forest Variation," discusses the regional and altitudinal influences on the forests, and examines particular habitats within the larger landscape. The dependency of forest composition, tree size and age, and the presence of a particular species upon the physical environment, is explained in detail.

Chapter Five, "Forest Processes," takes a closer look at the factors that determine forest composition. These include biological processes, such as tree germination or death, and physical processes, such as disturbance from flood or fire. The discussion provides some insight into the variance in age, species composition, and overall structure of the individual forest.

The sixth chapter, "The Forest Revisited," provides a summary of the first four chapters.

Forests of Paradise

The goal of this book has always been to extend beyond the realm of technical knowledge. It is my hope that with knowledge comes greater awareness and feeling for the magnificent trees and the forest community of which they are a part. Thus, as a preface to many of the sections of the book, I have departed from the technical discussion. These few sentences are written in italics at the beginning of selected sections and chapters. I have further explored this avenue in the sixth chapter, "Reflections". This chapter departs entirely from discussion and reflects my feelings and reactions to my experiences at Mount Rainier. For it is my belief that, in the end, our relationship with nature can be defined only partially from facts. These unique forests are neither uniform nor unchanging, but comprise complex entities driven by the interactions of many forces. Their beauty and grandeur far surpass any scientific discussion of their functionality, and it is my hope that with knowledge comes greater awareness.

The book concludes with three Appendices. Appendix A follows the format of a field guide, describing each species individually. Included here are some sizes of individual trees at the mountain; these were measured by the author. Appendix B provides a description of the reproductive organs of trees and their evolutionary history. Appendix C gives detailed descriptions of the forests along the two major Park highways, the Paradise and Sunrise roads. This appendix can be used as a road guide to explore the forests in more detail. Mileage and elevations, as well as descriptions of the forests, are given.

Although the book describes the forests at Mount Rainier, the scope extends beyond the Park boundaries to a much wider geographical area. The tree species and their forest associations are found throughout the western, and, in some cases, the eastern slope of the Cascades. The processes are common to other forests.

The mountain from Van Trump Park

One goal of this book is to inform the reader about the tree species that find suitable habitat at Mount Rainier. Although the discussion of the other plant species that comprise the forest community is minimized, recognition of associations of tree species and the many factors that determine forest composition is presented. The forests are much more than a group of trees. If they were only that, the forests at Mount Rainier would resemble plantations, those simple systems that lack both the diversity and the relationships found at the Park. It is my hope that the presentation of the tree species and their forests will provoke further interest. To a great extent, the trees define the physical and aesthetic environment at Mount Rainier National Park. They await discovery.

The Forest Visited

"A little more than 100 years ago, western Washington, then part of Oregon, was clothed with forests of giant trees from the Cascades to the Pacific and from the British line to the Columbia River. In these forests no trees had ever been cut, save for the buildings of the long-abandoned Spanish settlements of more than a half-century before, the buildings and grounds of the Puget Sound Agricultural Company at Nisqually, the tiny clearings of a few settlers on the Cowlitz, far to the southward of Tumwater, and cedars felled by Indians for the construction of their houses and canoes. It was almost the perfect, untouched, pathless wilderness, where trees grew from 200 to 300 feet in height, in places standing so close together that sunlight never reached the ground-often amid a tangled mass of fallen, rotting timber 30 to 50 feet deep." - Capt. N.A. McDougall, 1954

Two major tree groups comprise the forests of Mount Rainier. In terms of numbers and species the conifers are the most common. Populating the lowest to highest elevation forests, the conifers are gymnosperms, a term meaning "naked seed" in reference to the fact that their seeds do not have a protective coating. Conifers bear needle or needle-like appendages retained throughout the year on most species. The second group, the broad-leaf trees, are members of the more recently evolved angiosperm group. Angiosperms developed a new reproductive scheme, the flower. This organ utilizes animals, particularly the insects, for fertilization and distribution. The partnership has undoubtedly contributed to the diversification and wide distribution of angiosperms. The angiosperm species at the mountain are deciduous, meaning that they lose their leaves each year.

Subalpine forest

Among the most productive ecosystems on Earth, the Mount Rainier forests cover over seventy-five percent of the Park's area. There are at least sixteen coniferous and five deciduous tree species. The conifers are found throughout the entire altitudinal range of forested habitat, while most of the deciduous species are confined to elevations below 3500 feet. With increasing elevation, forest composition transitions from the large lowland species to smaller ones adapted to the highest regimes. Thus, elevation is one of the most important factors in forest composition.

The disturbance history of a site strongly influences forest age and composition. Disturbance parameters include fire, avalanche, mud flows, and wind. Localized physical parameters, such as slope of the land and angle with respect to the sun (the aspect), and climatic variations, also play a role. Biological factors that impact species distribution include dispersal mechanisms, disease infestations, shade tolerance, and reproductive strategies. Far from static entities, forests vary on short and long time scales in response to the interplay of the many physical and biological parameters.

A forest is much more than a community of tree species. Shrubs and herbs comprise an understory of varying diversity and age, and entire plant communities of species found only above the forest floor, thrive on tree trunks, limbs, twigs, and leafy appendages. Below, soils of varying age and composition provide sustenance to a multitude of invertebrates that constantly remake its contents. Vertebrates, too, make their home here, each in turn relying upon the provisions of a dynamic community. Woodpeckers excavate dead trees, while rodents cache seeds for sustenance through a long winter. A community of many members bound together in complex interdependency, the forest is an endlessly varying organism.

The forests at Mount Rainier share many features in common with other Northwest communities. Passing the Park boundaries on the main highways, the visitor enters regions of uncut forests. The deep forests here enhance the beauty of Mount Rainier which rises as an icy cone amidst a dark sea. In fact, in many ways they define the region as surely as the mountain itself; the views would certainly be altered if Mount Rainier was situated in a desert!

Subalpine forest mosaic

At higher elevations, the trees are much smaller than the lowlands. Near the limits of tree growth, the forest is a patchwork mosaic rather than a continuous community. The players change with location, both in terms of elevation and geographical position. Forest age is not determined by elevation, but is rather a function of various disturbance fac-

Rock-covered Carbon Glacier

tors, such as fire, flood, or avalanche. These impacts are common enough to prevent the development of many truly ancient forests; such forests are confined to small areas throughout the Park.

The ice ages, too, have influenced today's forests. During the glacial maximum thousands of years ago, Mount Rainier and the surrounding landscape were covered with thick ice, precluding forest establishment. At the edge of the great sheets the forest was composed of species found here today but of different importance to forest composition. As the glaciers retreated, species moved northward and upward in response to the altered landscape. Conditions favored some over others, however; for example, species adapted to the rocky and sandy substrate of the denuded land advanced into opened habitat. The incursion of other species was undoubtedly slowed by the infertile landscape. In more recent times, the advances and retreats the mountain's many active glaciers have continued to influence the forests.

Glacial moraine

Thus, the forests of Mount Rainier are neither uniform in composition nor static organisms. They are subject to a multitude of stresses, both ancient and modern. Analytical study of the contribution of these factors and forest processes is relatively recent, and knowledge is rudimentary. This serves as a warning, for manipulation, both intentional and inadvertent, based upon an elementary understanding poses both a great challenge and a great risk. It is within this tightly woven matrix of plants and animals that the beauty of the forests manifests itself. It is here also that the greatest danger lies.

The Volcano and Its Forests

Geologically speaking, the land is in its infancy. Dated at barely a half-million years ago, Mount Rainier's birth was a spasmodic and violent event. Spewing rocky deposits over terrain sculpted by more ancient outbursts, the internal forces of the young volcano altered the surrounding landscape and affected changes on distant lands. Volcanic outpourings built broad ridges, redefined river beds, filled valleys, and, inundating vast areas of vegetation, buried older and thicker soils, leaving a gravestone of rocky rubble. In brief geologic time, the young mountain changed the plant and animal communities inhabiting the old lowland plain. Conditions were harsh in the ravaged landscape, but new opportunities abounded in the vast areas of denuded terrain.

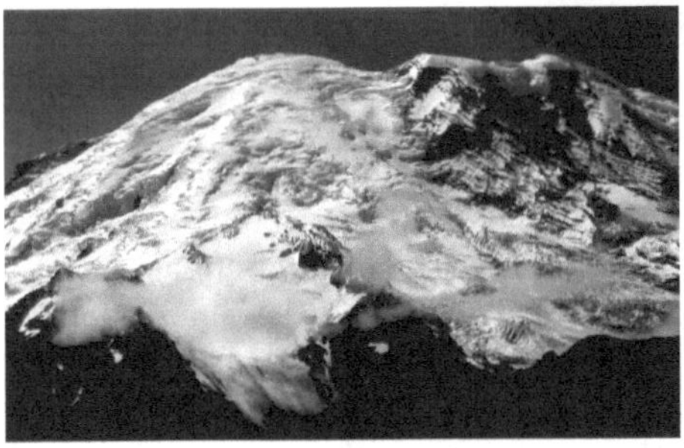

The Mountain from Third Burroughs Mountain

Volcanism provided the raw materials--ice sculpted it. The expanding mountain provided a higher elevation landscape upon which the glaciers, engorged by ice age conditions, imposed their own changes. Beneath the weight and power of moving ice, the elements of the volcano were stripped, ground down, and redeposited. Scouring whole mountainsides and carving basins where none previously existed, the rivers of ice transported much of the dislodged debris downslope, leaving long ridges of material (moraines) to mark their advance. Gray rivers carried away the rock ground and entombed within the glacial body, depositing their load in great, wide plains. As climatic conditions changed, receding glaciers uncovered vast beds of jumbled raw materials that became the substrate for invading plant life. Even in their diminished form today, glaciers contribute to wearing the mountain down.

Although ice-altered terrain is dramatic, water in its liquid form is an equally important landscaper. Through periods of quiet stability and times of dynamic change, the ocean's waters, evaporated by the sun and transported by the wind, have re-worked the land. Sourced in rains or melting snows, tiny trickles and torrential currents alike transform into powerful grinding forces. High altitude creeks wind through deep channels cut along the base of steep slopes and fall in great cascades over giant volcanic escarpments. Torrential rivers transport material by the ton, dropping great loads along their changing course.

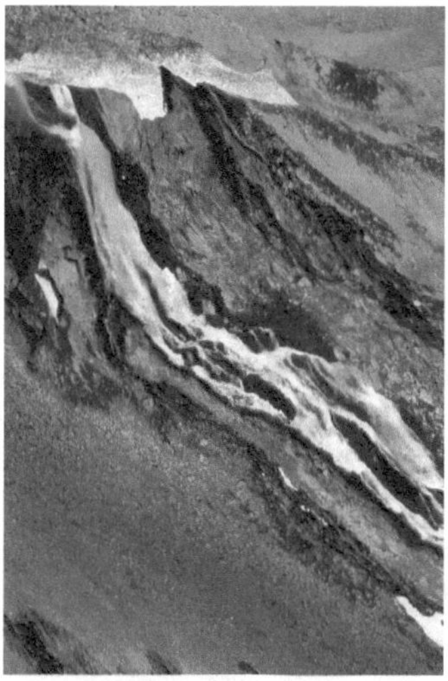

Waterfall over glaciated terrain

Given enough time, even the smallest stream slices and parts the earth like a knife, exposing the naked rock.

In a restless slumber today, Mount Rainier continues to influence the plant and animal communities that make the mountain their home. The volcanic embers are not dead. Hot, chemically enriched waters contribute to the formation of unique habitats, changes in glaciers open and close the terrain to pioneering species, and gigantic ice and mud slides triggered by invisible volcanic activity decimate entire landscapes, including those that are home to humans. Small climatic variations alter the forests on an observable time scale, opening landscapes to pioneering species, while engorging the rivers which sweep the valleys, snatching land once home to plant and animal alike. Influenced by factors far from its boundaries, the variability of the environment at Mount Rainier ensures a constant challenge for the forest members.

The Tree Species

The coniferous and deciduous tree species at Mount Rainier are members of several plant families. Among the conifers, the pine family (Pinaceae) includes the true firs (the *Abies* genus), the hemlocks (*Tsuga*), spruces (*Picea*), pines (*Pinus*), and "pseudo"-hemlock (*Pseudotsuga*). The cypress (family Cupressaceae) genera comprise what are labeled as "cedars," including Western redcedar (*Thuja*), and Alaska cedar (*Chameocyparis*). Pacific yew (*Taxus*), a member of the Taxaceae family, is perhaps best known for the cancer treatment derivative derived from its bark. The leaf-bearing species include the maple (Aceraceae) family, the willows (Salicaceae), the alders (Betulaceae), and a rose, Bitter cherry (family Rosaceae). Typically, the conifers live longer and achieve much greater size than their leaf-bearing associates, but there are two large deciduous species, the Black cottonwood and Bigleaf maple.

Pacific yew

There are four true fir (*Abies*) species, including Subalpine fir, Pacific silver fir, Grand fir, and Noble fir. The common Douglas-fir is the only Northwest representative of the *Pseudotsuga* genus, meaning false hemlock. The pines include Western white pine, Lodgepole pine, Whitebark pine, and Ponderosa pine, while the spruces, Sitka and Engelmann, are low and high altitude species respectively. Pacific yew is a unique coniferous species which bears a red berry-like structure rather than the typical cone. Vine maple and Bigleaf maple are abundant deciduous species belonging to the *Acer* genus (with the rare Douglas maple found on occasion), while Red alder and Slide alder (*Alnus*) occupy elevations from the lowest to highest regimes. Black cottonwood forms massive communities along rivers, and the related willows inhabit the wettest areas. Finally, the uncommon Bitter cherry is easily recognized in the springtime when its white flowers add a touch of color to the lowland forest.

With their symmetrical, pyramidal configurations, the coniferous species may seem difficult to distinguish from one another. There are observable differences, however; for example, the four pine species bear their needles in bunches, while fir and spruce needles are arranged in rows along the twigs. The cedars are unique with their nearly vertical fern-like branches that give the two species a droopy appearance. The bark of the conifers differs, too, from the thick, grooved, red-and-gray bark of old Douglas-fir trees, to the long shredded bark of the cedars and the small plates of Western white pine. Old Noble firs are covered in bark with a maroon tint, and Pacific yew barely conceals a rich, reddish trunk beneath flaky bark.

Bigleaf maple

The deciduous species are most easily distinguished from the conifers by their leaves, and a more rounded or shrubby appearance. Both of the common maples – Bigleaf maple and Vine maple – bear lobed leaves, but tree and leaf size easily distinguish these species. Slide alder is characteristically shrublike with multiple trunks, and bears leaves with double, pointed "teeth," while Red alder adopts a more conventional tree form and supports leaves with single teeth. It is Slide alder that is commonly found in higher elevation avalanche chutes, while Red alder inhabits lowland wet sites.

Noble fir cones

The large scale features are not the only characteristics that separate the coniferous tree species. Each conifer bears distinctive cones, ranging from the small nutlike appendages of Western hemlock to the eight-inch Western white pine cone. The true firs place their cones far out of reach above the forest floor, with a size ranging from the two-inch Grand fir cones, to the large and heavy Noble fir cones. The differences in seeds are much more difficult to observe. They are typically small; for example, it takes 14,600 Noble fir seeds to weigh in at a single pound, which is large by comparison to the hemlocks.

There are many hazards to overcome before attaining a stature as great as an old Noble fir. Seedling branches may be snapped or bent by the heavy snows, roots may drown in excessive moisture, while competition for sunlight, nutrients, and water is often intense. Protection is provided by other trees and the snow pack itself, but thousands of seeds fail to germinate, and seedling death rate is high as well. For example, ninety percent of Subalpine fir seedlings may perish in a few seasons, while Grand fir, although more successful, typically is subject to a forty percent seedling death rate within two years.

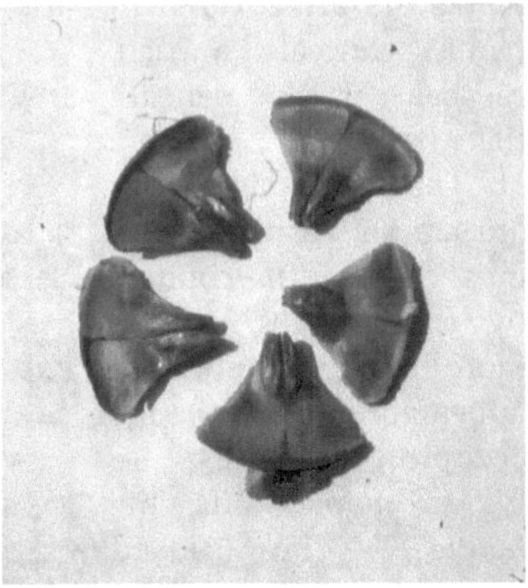

Noble fir seeds and scales

In spite of their apparent vulnerability, however, the trees possess a few tricks of their own. Young Mountain hemlocks lie sheltered beneath the snows, waiting for release from the burden, their tops noisily snapping as the snow melts. Pacific silver fir seeds germinate as quickly as possible in the spring, sometimes on top of the lingering snow pack that surrounds the parent tree. Their moisture requirements give the appearance of being caught between two opposing forces—the need for moist soils during the growing season, and the dangers of germinating before the snow pack dissipates. Many species increase the chances for germination by producing large quantities of seeds, most of them falling near the parent tree. Some, however, such as the Whitebark pine, produce fewer seeds and depend on other means of dispersal to carry the seeds to an advantageous location.

Adapted to the demands of a variable environment, the tree species of Mount Rainier differ in form and function. They flourish upon a range of soil types, and under a variety of climatic conditions. Their establishment is hazardous, their continuance threatened by drought, flood, fire, avalanche, wind, and snow. The realities of their environment have produced long-lived robust species. To a great extent, they define the beauty of the mountain.

The Species Associations

The forests at Mount Rainier are not uniform in composition, but consist of associations of species. From an altitude of less than 2,000 feet, to the limits of tree growth at over 7,000 feet, a wide range of conditions define the habitats for each distinctive forest community. Strongly but not exclusively dependent on elevation, these associations define the divisions of the low, middle, and high (subalpine) altitude forests.

Each association is characterized by a set of tree species and other plants. The latter are part of the "understory" community; that is, those plants that grow beneath the forest canopy. These include the shrubs, flowering plants, mosses, ferns, and grasses. The relative abundance of the various species as well as the size and age are strongly determined by the time since the forest was established. Frequent "disturbance" factor impacts, such as fire and wind, ensure a variation in forest age from site to site. Disturbance favors certain species; as temporary residents of the landscape these are the "seral" species of young communities. Typically adapted to an open environment, seral species may require direct sun for establishment. As the forest ages, these pioneers will be replaced by other species. But their continuance throughout the region is guaranteed by the frequency of disturbance events.

Young trees — streamside

As time passes without a major disturbance, species are added or lost to the community, and the abundance of any particular species may change. For example, a pioneering species that requires direct sun for establishment will not regenerate as the forest ages. In its place, species more adapted to shady conditions will begin to dominate. If the pioneering species is long-lived, however, it will maintain a presence, although in fewer numbers. An excellent example of this type of pioneering species is Douglas-fir. Long-lived, to a thousand years or more, Douglas-fir does not regenerate in an aging forest, but giant members of this species may overtop the forest canopy, amidst the smaller, shade tolerant associates.

Unlike a seral species, such as Douglas-fir, the new inheritors of the forest will replace themselves, and a community of multiple-aged members will form a "climax" forest. This type of forest is conceptually one in equilibrium with the environment, where changes in composition are minor compared to the overall structure and duration of the forest. Western hemlock is the common lowland climax species; in such a community, hemlocks may vary from seedling size to large members established early in the forest's history.

The concept of seral and climax forests implies that associations of species exist under temporal conditions. Other physical parameters contribute to the presence of a particular species. Glacial history, rainfall, soils—all play a role in determining the composition. At Mount Rainier, elevation is a very important factor. Regardless of the age of a particular community (and hence whether it is dominated by pioneering or climax species), the forest communities may be specified as "low," "middle," and "high" (subalpine) forests in acknowledgment of the altitudinal influence. Each of these regions is home to a particular set of species.

Western hemlock

The lowland forest occupies a range of approximately 2,000 to 3,000 feet at Mount Rainier. Coniferous species include Douglas-fir, Western hemlock, and Western redcedar, and, less frequently, Pacific yew, and Grand fir. Red alder, a common deciduous disturbance species, populates stream banks and wetlands at the lower elevations; other typical leaf-bearing species in this habitat include Bigleaf maple, Black cottonwood, and Bitter cherry. Red alder is a pioneering species; as the forest ages, Western hemlock and Western redcedar will replace the alder. Douglas-fir and Western hemlock are common coniferous associate species in the lowland forest, with Western redcedar as a lesser component. As this type of forest ages, Western hemlock replaces the pioneering Douglas-fir as the climax species. Old forests typically consist of a few gigantic Douglas-firs, and a multiple-aged community of Western hemlock.

Although not common, Sitka spruce and Grand fir commingle in the sandy regions along select glacial rivers that thread through wide, lowland valleys. Western redcedar is a lesser component of this forest, while the deciduous members include those common at other lowland sites. Western redcedar is often found in the wet habitats preferred by deciduous species such as Black cottonwood and Bigleaf maple. Throughout the lowland region, Vine maple plays an occasional role as an understory species.

Pacific willow

The middle altitude forest begins at an elevation of approximately 3,000 feet, and is composed of a new group of deciduous and coniferous species. These include Pacific silver fir, the most common species at the Park; Alaska cedar, a relative of Western redcedar; Western white pine, and Noble fir. Although present, leaf-bearing trees are both smaller and fewer in number. Along moist meadow edges and stream banks, willows and Slide alder grow, while, in association with the alder, Vine maple inhabits the rocky and steep terrain more common at this elevation.

Subalpine fir

The subalpine forest begins at an altitude of approximately 5,000 feet and is composed of those tree species with ranges that extend nearly to the limits of plant growth. Mountain hemlock, Subalpine fir, Engelmann spruce, and, at the loftiest altitudes, Whitebark pine, constitute the subalpine forest. Within this domain, the species are restricted; for example, Pacific silver fir extends its middle altitude range into this habitat, but does not grow to as great an elevation as Subalpine fir. Other species, such as Whitebark pine and Engelmann spruce, are found only on the north and east sides of the mountain, whereas Subalpine fir grows throughout the entire subalpine forest. Alaska cedar is often found on the wet sites also attractive to Engelmann spruce.

Local and regional variation in forest composition and form is influenced by many factors. For example, the drought tolerant Whitebark pine grows on porous volcanic soils where it may adopt a stunted form. But in more hospitable sites, this species assumes a more conventional tree form, albeit with multiple trunks and a reduced height. Subalpine fir occasionally grows in cold air drainage sites in the middle altitude forest, where it may reach substantial size. Western hemlock extends its range to altitudes of more than 4,000 feet on well-drained sites, and Western redcedar, more typically a lowland species, also grows in the middle altitude zone. This species in particular gives the impression of an opportunistic pioneer, thriving as individuals or in small communities in a variety of habitats.

Engelmann spruce

A species restricted in range to the north and east sides of the mountain, the cold tolerant Engelmann spruce grows from the middle elevation forest to the highest limits of tree growth where its most common associates include Subalpine fir and Alaska cedar. This wide range makes the geographical restriction of this species puzzling; possible explanations include the proximity of the north and east slopes of the mountain to the eastern side of the Cascades where Engelmann spruce is common. Other factors may include the drier climate on this sheltered side of the mountain, and the peculiarities of glacial history. Engelmann spruce may live several hundred years once established, but seedling viability is marginal, with many factors affecting survival. These include but are not limited to, location, variance in climate, the presence or absence of other trees, and the quantity and distribution of snow.

Extent in range is one example of the spatial variability in tree associations. Engelmann spruce provides an example, as this species range was more extensive during the recent ice age. Even the most established forests are subjected to local variation, where opportunity may be restricted but not eliminated. Interacting with one another, and adapted to specific conditions, associations may change with time or place. An example of this is evident on the north side of the mountain in the recently de-glaciated terrain of the Winthrop Glacier. Engelmann spruce and Lodgepole pine are pioneer species here, but in the event of changes to the glacier, this association may disappear altogether. On the other hand, a change in local conditions may promote the establishment of another species within this group.

Thus, continually subject to changing conditions, a particular site may play host to more than one association of species. The invasion of another species adapted to a slight variation in conditions may also alter the community. Some associations are short-lived at one location, but maintain an ongoing presence at the mountain in similar conditions at other sites. Others endure for centuries, with individuals replaced by their own species.

Winthrop glacier — Engelmann spruce

Tree Types

Overview

Gymnosperms and angiosperms are members of the Tracheophyta plant phylum. One of five divisions of the plant kingdom, in terms of species the tracheophytes have diversified into the most widespread continental plants. Evolving from a common ancestor with the bryophytes, another land phylum, the stunning success of the tracheophytes may be attributed to two unique features. Tracheophytes utilize vessels for food and water transport; this mechanism removes the dependency on direct absorption of these elements through surface tissues. These distinct vascular systems, the phloem and the xylem, conduct the products of photosynthesis and water respectively. This efficient method has facilitated the evolution of trees, the tallest of living beings, and has undoubtedly promoted the extension of their range into the most varied habitats.

Pollen cones

Lacking vessels, the bryophytes must absorb food and water through their surfaces, a restriction that limits both size and habitat. Bryophytes also depend on standing water for reproduction, while tracheophyte fertilization is free of this mechanism. In modern tracheophytes, germination takes place from the seed, a product of the union of the male component, carried in pollen grains, and the female structure. Packed with a food source for germination, seeds employ several means for dispersal, enabling a wider range for the tracheophytes .

Thus, only the tracheophytes have solved the dual problems of reproduction without water transport, and efficient delivery of food and water to tissues. They are not the only vascular members of the plant kingdom, but in them the system is most specialized, while their leaves are more complex than other plants. Although the bryophytes are represented today, it is the tracheophytes that have adapted most successfully to the challenges of the land.

Gymnosperms

Perched 200 feet above the ground, their needled crowns seem to touch the sky. In their canopy lives a community that never descends. Reduced to a storm-battered shrub, they survive at the limits of life. Of ancient lineage, they live for centuries. They are the gymnosperms.

Gymnosperms appeared around 250 million years ago during a time when the Earth's climate cooled and dried. The word gymnosperm means "naked seed," a reference to the fact that the seed is encapsulated within an ovule only. The gymnosperms are members of a tracheophyte subphylum, the Pterophtina, a group that includes four classes of gymnosperms—the cycads, gingkos, gnetinae, and the conifers, as well as the ferns and angiosperms. Conifers are the largest and most common gymnosperms, and it is members of this class that dominate the forest landscape at Mount Rainier. The tall, dark green, familiar trees are conifers. They include the common Douglas-fir and the more rare Pacific yew, the long lived Western redcedar and Alaska cedar, and the versatile Subalpine fir.

Pacific silver fir needles

Conifers are widespread, with a range from the warm subtropics to the subarctic extremes. This cosmopolitan characteristic is aided by the coniferous needle whose shape and composition indicate the significance of the development during dry and cool times. The needle is long and narrow and covered with a waxy cuticle. The cuticle cover and the small "stomatal" openings on the needles, used for respiration, facilitate water conservation. In most conifers, the needle is evergreen, held throughout the year and thus available for every possible photosynthetic opportunity. Photosynthesis, the process that utilizes the sun's energy as a first step in the production of food for the tree, occurs within the green pigmented needle cells. Varying from less than half-an-inch to more than half-a-foot, the versatile needle sustains life.

Top of Douglas-fir

In mature forests, the canopies of ancient trees tower above their surroundings. Such a height strategy provides access to unblocked sunlight, a clear advantage for the gigantic trees. But size of this magnitude poses two major problems--the great strength required to support the massive trunks, and the transporting of water and food throughout their length. As with all plants, gymnosperm cells are in large part structured from cellulose, a strongly bonded sugar molecule. It is the strength of cellulose that makes size possible. In addition, tracheophyte vessels are able to transport water from roots to crown and food throughout the entire tree.

Not all gymnosperms attain the height of the Northwest giants, however. The combination of an abundance of moisture in the form of rain and snow, and the availability of water during the dry summer months certainly plays a role in their immensity. With relatives of world wide distribution, in the forests of Mount Rainier the gymnosperm families have achieved great success.

Angiosperms

Along the streams and rivers, bare-boned branches display their angularity against the stark grayness of winter.

Bigleaf maple — an angio-

Like the gymnosperms, angiosperms are tracheophytes, species with a vascular food and water transport system. They arrived much later on the scene, having evolved around 100 million years ago during a warm and mild period in Earth's history. Probably originating from an ancient gymnosperm, angiosperms may have spread from infertile uplands into the warm and crowded lowlands where they were destined to become the dominant plants. It is the angiosperms that incorporated a new structure into their reproductive strategy—the flower. Flowers enabled this quickly diversifying group to take advantage insects, a willing and available fertilization mechanism.

Bigleaf maple flowers

In contrast to the coniferous gymnosperms, which retain their needle coat throughout the year, the angiosperms bear less resilient leaves that typically fall as temperatures drop. Thus, unlike the cold resistant gymnosperms which may photosynthesize any time the temperature is sufficiently high, angiosperms forgo photosynthesis altogether until the warmer days of spring.

At Mount Rainier, most angiosperms are less adapted to cold than gymnosperms and are thus confined to lowland habitats. Large species in particular are unable to survive in the higher elevations; Black cottonwood, for example, is suffused with water throughout its trunk, making it very susceptible to breakage during freezing and thawing cycles.

Not only are the larger angiosperms absent from higher elevations, but the group as a whole is represented by fewer species at Mount Rainier. Along the lowland glacial rivers, Black cottonwood, Bigleaf maple, and Red alder grow abundantly, but only the small, flexible willows, Slide alder, and Vine maple invade the higher regimes. There they are relegated to wetland or avalanche conditions, but even these small trees do not reach the highest limits of tree growth. That environment is reserved for the conifers.

Black cottonwood

In contrast to the starkness of their winter appearance, the angiosperms brighten the springtime forest. Against a backdrop of the dark conifers, pale buds swell and open in a profusion of tiny leaflets, transforming the stark trees. Branches sway in the breeze, minute flowers mature to produce windborne seeds, sun-turned leaves ripple, and even the age-grooved trunks seem younger when topped with their green canopy.

Vine maple autumn leaves

The beauty and showiness of this leafy cover perhaps masks the seriousness of its intent. Northwest summers are brief and occasionally cool and wet. Constrained to a four or five month growing season, the leaves must capture the available light energy during that period. It is the leaves that regulate water resources, and serve as receptacles and transformation site for atmospheric carbon dioxide. Without sufficient chlorophyll present in the green leaves, the deciduous angiosperms cannot survive.

In the fall, angiosperm leaves are transformed from green to yellows, browns, and reds. With a blazing farewell to summer, Vine maple contributes its red leaves to the forest floor. There, in the unseen microscopic world, the decaying leaves supports the continuance of the angiosperms and their distant relatives, the old gymnosperms.

Tree Parts

Soil

The old road wound up a small hill, flattened slightly, and passed onto a grated bridge. Upstream, rocky walls constricted the gray river. Cutting through this boundary, the road exposed the volcanic rock. Hard and sterile, the terrain could not have appeared more uninviting to life. But amidst a small plant community of lichens and mosses, a tenacious conifer gripped the rock, creating soil from the hard minerals.

With their blankets of solidified rock, fresh volcanic landscapes provide inorganic material for the soil-building processes. Other processes, too, including those that alter the original rock substrate, contribute parent material. As a product of physical and biological factors, soil evolves in partnership with the plant and animal community. Soil may be young or old, richly organic or mineral based, deep or shallow, airy or dense. The nutrient content of soil provides both opportunity and limitation to the biological community that depends upon it, while its variation in time and space creates a changing scenario for plants and animals alike.

Fungi

Supporting a diverse community of plants, soils are formed from both destructive and constructive processes. The breakdown of the regolith, an unconsolidated material that forms the building blocks for soil, is carried out by physical forces such as weathering from rainfall, and biological interactions, such as the disintegration of plant fiber by small organisms. Plant roots extend into this material, breaking it apart and contributing organic matter to the mineral base. Constructive processes include chemical reactions that result in new minerals, such as those found in clay. From these interactions between physical and biological factors, soils of varying content are formed. From the tiniest bacteria, to worms and insects, to the vital mycorrhizal fungi bound to the tree roots, to the roots themselves, a host of organisms make soil their home. All are important components for soil maintenance.

Most soils are not uniform in composition but are distinctively layered in a manner that reflects their origins. For a typical soil, these layers are given letter names: O, A, B, and C, and each is divided further into horizons. The top layer (closest to the surface) is the "O"; this is the layer that contains the most organic material from dead plant and animal surface sources. Below this layer and often laying close to the surface is the "A" horizon, a region where minerals leach freely downward for deposition in the "B" horizon. This layer is a region of maximum accumulation of clays, oxides, and other minerals deposited from above or formed within the layer. The "C" layer is closest to the bedrock, and is least affected by the processes that form the soil above. The thickness of these layers is variable, and, in some cases, a horizon may be missing altogether or the soil might not be layered at all. Human processes, such as agriculture, may impact the layers and their composition, as does the parent material, the geologic history, and other natural forces to which the soil formation process has been subjected.

Historically, a naming convention for soils was developed that described the distinctiveness of the soils. The names take into account the variance in soil origins and composition, including the materials present in the various layers. This classification system is based on a four part division of the soils; these are order, suborder, great group, and subgroup. There are ten soil orders that reflect the age and origin of the soils. In the Northwest, soils that support the coniferous forests are usually spodosols, which means wood ash. Typically light colored, these are soils which are formed by podzolic (acid leaching) processes. Although mineral in nature, spodosols usually have a layer of organic accumulation. Thus, these soils are both acidic and subject to downward leaching through the soil layers; they are most commonly found in cold and temperate climates and are formed from the breakdown of organic materials by fungi.

Glacial moraine along Puyallup River

At Mount Rainier, mixtures of volcanic deposits from the mountain or surrounding volcanoes, and glacially deposited material, constitute the parent material for most soils. Much of the soil is coarse, unconsolidated material that drains readily. Deposition along stream and river valleys results in alluvial soils, well-drained mixtures composed of gravely materials, and moraines of glacial till that lend variety in particle size. Generally, these soils are young and potentially fertile, although less likely to serve as sites for nutrient exchange owing to their crystalline nature.

The processes that are vital to plant life are strongly determined by the clay and humus content of the soil. Nutrient exchanges and chemical reactions are determined in large measure by the size, composition, and presence of these two elements. Clays are laminated, crystalline structures that exist in the colloidal state and are negatively charged, a state that gives clay the capacity to serve as a site for chemical processes. Clay particles are hydrated materials that provide large amounts of external and, in many cases, internal, surface structure for active ion exchange. Thus, clay is both a repository of chemicals vital to plant life, and an exchange and reactive site for those chemicals.

Organic humus plays an important role by providing the molecules and their constituents in a form either usable by living plants or available for further processing by soil organisms. The organisms themselves contribute organic matter upon their death; many of these are plants of one form or another, but a multitude of animals are also present within the soil.

One of the most important of the soil plant organisms is the mycorrhizae. These are fungi that invade the roots of plants, forming a symbiotic relationship with the host and playing an important role in breaking down organic material. This invasion provides benefits to both plant and its associated fungi. Plants which are host to mycorrhizae generally are more robust than those lacking the fungi; this is due primarily to the increased uptake of nutrients by the fungi. In return, the mycorrhizae receive food in the form of carbohydrates bound within the roots. Mycorrhizae are very important in the transformation of atmospheric nitrogen to a form which may be further processed by micro-organisms within the soil. They may also provide protection to the plant roots against invasive attack by parasites.

Plants are dependent upon many nutrients, some in appreciable quantity, and others in minute amounts. Not all elements are present in usable form, and the presence of a mechanism, either biological or physical, to alter these elements is as important as the presence of the chemicals themselves. Some elements, such as potassium and calcium, are often readily available, whereas phosphorus and nitrogen are not so easily obtained. Organic nitrogen supplied by plants must be converted to an inorganic compound by soil bacteria. Roots are particularly important in providing organic material; this substance must be broken down by organisms to usable form. At Mount Rainier, the contribution of plant parts to the soil building process is especially important due to the thin and sandy nature of the substrate.

Glacial recession along Nisqually River

Factors that influence soil composition include climate, parent materials, vegetation, and time. In the Northwest, soils have been strongly affected by the ice sheets of the most recent Pleistocene glaciations. Continental in extent and episodic in nature, the impact of these glacial irruptions on soils is difficult to overstate. Enormous quantities of soil were moved and re-deposited; whole areas stripped bare, plains of outwash deposited, sides of mountains sliced and lumped onto long ridges. Rocks and sand coexist with silt and clayey deposits; even the wind played a part in moving materials from one region to another. The soils at the mountain strongly reflect this glacial heritage.

Thus, soils vary from deep, dark and ancient, to shallow, sandy, and stony. Soil is the supporting substrate for the life that sets on its surface and penetrates its depths. The youth and variation in the soils at Mount Rainier provide insight into the robustness of the plants, and, in particular, the trees. In spite of the apparent infertility and instability, trees thrive, forming dense forests in the lower and middle elevations. However, both forest composition and individual tree size are influenced by the interplay between plants and soil forming processes. From rich river bottomlands to scraped ridges, the trees also contribute to the modification of the soils that supports them

Roots

The narrow trail carved a thin line through the old forest. Here giant trees waved their crowns against a distant sky and crowded the path. Disappearing into a bend, the trail emerged onto a narrow plateau above a stream and widened, to pass through a jumbled terrain. The scene was one of destruction, as twigs, needled branches, and pieces of trunks lay scattered about. A gaping hole free of debris had swallowed a piece of trail. Ripped from its moorings, a giant tree lay nearby, its roots waving in the air, forming a flattened structure many feet wide. Rocks hung enmeshed in entwining roots that ranged from barely more than a thin string to thumb-sized twigs. Within the immense hole the earth was churned, torn by the fall, and exposed to the air. In dying the tree provided opportunity in a naked gap ripe for habitation.

Tree roots laid bare

The falling of a tree is an important process in forest ecology. Most tree species develop extensive lateral root systems which provide habitat for a myriad of creatures below the surface, and new niches for a different community of plants and animals when extracted from the ground. The subterranean life that thrives amongst live tree roots is exposed to threatening conditions when the tree is uprooted. Constructive processes that increased and nourished the living roots are now replaced with decomposing organisms.

These new organisms break down the roots to organic compounds available to other plants and animals. The exposed soil in the mass of roots provides a whole new habitat, a rich source for non-decomposing plants.

The amount of lateral and vertical root development, as well as the density of the roots, is dependent not only on the species of tree, but on the particular site which the tree occupies. Some species typically develop long taproots that penetrate the soil deeply. These include the pines and others adapted to dry or rocky habitats. Ponderosa pine develops such a root, as does Whitebark pine, a species that grows on rocky sites at high elevations. Other species develop deep roots in dry habitats, but may forgo the taproot in moist conditions. Grand fir owes in part its capability to inhabit moist and dry sites by developing a taproot in the drier conditions, while growing a more extensive lateral system in wetter locations. Pacific silver fir often develops a taproot which penetrates compact soils; Douglas-fir, which may inhabit relatively dry sites, also develops deep roots.

Species adapted to wetter conditions lack the deep roots, but develop extensive lateral root systems. Although these species are prone to "windthrow" (that is, a tendency to be uprooted by the wind), the adaptation is a practical one, for there is no need to penetrate deeply if moisture is readily available. And surface root systems can take advantage of the fertility of the top organic layer in areas of thin soils. Western hemlock confines its roots to this rich surface; Western redcedar grows more deeply than the hemlock, but its roots are also predominantly confined to the surface region. Even Western white pine roots are most extensive in the top foot of soil; this species may spread over 20 feet laterally and develop a system of fine, fibrous roots. Sitka spruce develops a flat, plate-like system, or a deep and spreading set of roots. The related Engelmann spruce may also develop extensive lateral roots; this species of wet environments also has a tendency towards windthrow.

Flood damage

Many species may develop "adventitious" roots.

These are new roots that grow in surface material deposited after the initial establishment of the parent tree. This mechanism is particularly important where floods deposit silt layers around a tree, or in regions where volcanic deposits bury a portion of the trunk. Since roots require aeration, a tree may suffocate if the root system is buried; thus, the ability to grow these new roots in the deposit can spell the difference between life and death.

In many species, the roots are home to symbiotic mycorrhizae, fungi that are a conduit for nutrients. Mycorrhizae consist of two distinct types. The first is the ectotrophic group which penetrates between root cells; the second consists of the endotrophic group which actually enters the outermost layer of cells (the epidermal cells) and the tissue immediately beneath (the cortex root cells).

Root cells exude carbon dioxide, a compound that speeds up nutrient interchange. The immediate root zone may support ten times as many organisms as areas away from the roots, and absorption of water by root hairs creates a tension which draws more water for the tree. Excretion of organic compounds provides food and energy for organisms, and the roots themselves contribute enormous amounts of organic matter to the soil building process.

***Russula* — mycorrhizal fungus**

Roots come under attack from a variety of root diseases, including root rotting fungi, organisms that may be of particular concern in fire ravaged regions. Heavy infection of roots often results in a reduction in the tree's growth rate, a yellowing of foliage, or an exuding of dying roots. Roots must have sufficient oxygen to remain healthy; in cases where flooding occurs, the root system may cease to function altogether, spelling death for the tree.

A vital organism, roots serve as intermediary between the tree and the soil. They are the means by which the tree acquires water and nutrients. Extending myriad arms into the substrate, roots entwine small rocks, break apart whole slabs, and participate in soil making. They are the lifeline of the tree and the tiny organisms that live beneath the surface. They create symbiotic relationships beneficial to both the tree and the fungi itself. Beneath the surface, roots live amidst millions of creatures in a thriving community of mutual dependence. Popped to the surface by a windstorm, roots expose their waving tendrils to the air. In dying they are given over to a new purpose for the forest community. Continuously altering the, in life their successful functioning promotes a healthy, vigorous tree.

Bark

Hastily pasted on with no apparent plan, the thin bark seemed to preclude the necessity for order.

With a potential life span of 2,000 years, Alaska cedar is the longest-lived tree species at Mount Rainier. Reddish-gray and very thin on saplings, with age the color of the cedar's bark fades to light gray and is hung in long shaggy strips, giving the tree a disheveled look. Bark is such an important defense against disease, insects, and fire, it might be expected that to achieve such longevity Alaska cedar would support a thick covering. But although it is a vital organ, the advantages of thickness must be offset against the energy expenditure to create that resource. Since fire is infrequent in the moist habitat preferred by Alaska cedar, a thick covering is required only on rare occasions. The potential threat is, in some sense, offset against the necessity of protection, and thus this species supports bark only a half-inch thick.

Alaska cedar bark

Alaska cedar employs other defenses against such life-threatening attacks. Within the thin cedar bark, an arsenal of chemicals serve as a defense in the moist environments where attack by fungi is a continual threat. The presence of these toxic chemicals is undoubtedly an important contributor to longevity. Similarly cloaked with a thin bark, the related and long-lived Western redcedar also employs a chemical defense against invasion.

At the other extreme, Douglas-fir may support bark up to 12 inches thick. In terms of numbers perhaps the most successful conifer in the Northwest, the cosmopolitan Douglas-fir can live a thousand years, a lengthy life span at least in part attributable to the protective bark. Old trees growing in habitats subjected to frequent fires are often Douglas-fir, as this species is best equipped to survive the periodic conflagrations. Young Douglas-fir are covered with a muted gray bark, but with age, the bark thickens and develops deep furrows, varying in color from gray to reddish-brown. In cross-section, the grooves appear to be composed of wavy layers of tan, gray, and red, giving a warm appearance to the massive tree. The thick bark provides protection, and, on inspection, many old Douglas-fir reveal black slashes, scars of fiery battles fought and won.

Fire-scarred Douglas-fir

Unlike Douglas-fir, most coniferous species support relatively thin bark. Sitka spruce and Engelmann spruce are species of wet habitat and, as with Alaska cedar, appear to have little need for a thick covering. Western hemlock grows in a variety of habitats, but it, too, supports a thin bark. The bark of the true firs thickens with age, but these shorter-lived species are much less protected against fire and disease than Douglas-fir.

Most of the deciduous species in the Park grow in wet habitats, and are covered with thin bark. The largest deciduous species in the Northwest, Black cottonwood, bears a distinctive, fissured gray bark. Although hard and protective of the soft wood inside, the bark cannot prevent the invasion of fungi nor the frequent breakage to which this species is prone. Mature Red alder bark is often brightened with a whitish-colored lichen, while Bigleaf maple bark readily supports a rich community of mosses and ferns.

Black cottonwood bark

With a little practice, mature trees may be identified by their distinctive bark. Young trees are more difficult, as they are often uniformly gray in appearance, and although the true firs are blistered with resin pockets, other species may also exhibit this phenomenon. The bark of old true firs is gray and thickened. The two spruce species are covered with maroon-colored, cookie-cutter bark pieces. Mature Western hemlock bark is grooved with a brownish cast; it may be confused with Douglas-fir, but the grooves of the hemlock are more shallow and closer together. Mountain hemlock bark becomes dark and furrowed with age.

Perhaps the most easily recognized species by its bark is Douglas-fir. The contrast of the black and red, deeply furrowed bark imparts a beauty worthy of its' reputation. Where a giant has fallen across the trail and been subsequently cut for clearing, it is worthwhile to examine the slice. The bark may appear thin in contrast to the inner core, but the size of the tree and the presence of black fire scars are indicative of its protective function. Having yielded to internal attack from disease and insect, or perhaps lost the battle with fire or wind, the fir displays the thick covering that imparted the gift of longevity.

Branches

Through a screen of conifers the sinuous river twists through a forest of Red alder. Soon winter will be a memory, as the bare branches disappear within a curtain of leaves.

In late spring, the ends of conifer branches are brightened with the lime-green of new growth. Depending on the species and location, these soft new needles are variable in length. Even the prickly spruce needles, which resemble the distantly related Douglas-fir, are less sharp when young. More subtle than the deciduous trees, the conifers are quick to respond to the warm days.

The bare branches of deciduous trees are transformed with a burst of color and vigor. Within a few weeks six-inch long maroon branches of Bigleaf maple are tipped with giant leaves, while Vine maple bears smaller, star-shaped versions. New alder leaves unfold slowly, covering the gray branches, and the angular cottonwoods are softened with a profusion of their heart-shaped leaves. The importance of these leaf-supporting branches then becomes apparent, as it is the leaves that are the site of the photosynthetic process.

Bigleaf maple branches

Individually, branches may be unimpressive, but collectively they define the configuration of a species. Without them, a tree is just a pole, and a dead one at that. Covered with branches bearing thousands of needles or leaves, a tree adopts a unique appearance that both encompasses and surpasses the contribution of each stem. Their common task is masked in the uniqueness of the species they define.

Like the most vigorous conifer, the deciduous Black cottonwood thrusts a giant trunk skyward, only to bow to its heritage by developing an immense crown of many-angled branches. The cottonwoods tower above the swept up Red alders, while the Bigleaf maples are no less impressive with their wide crowns. Conifers, too, may be distinguished by the configuration of their branches; young Douglas-fir trees bear sweeping arcs which resemble a child's drawing, but upon maturing the long branches develop into multiple lengths that give the tree an irregular form. Noble fir branches are accessible on saplings, but old trees carry the lowest branch so far above reach that it is difficult to determine how this species supports sufficient surface area for photosynthesis. Filmy hairs cover the tan-colored twigs of Engelmann spruce, but the related Sitka spruce bears hairless branches. And Pacific yew seems to acknowledge its understory status by adopting a loose form, nearly twiggy form.

Black cottonwood

A single branch may provide a distinguishing field mark. The whip-like top of the Western hemlock is created by a protruding branch that sets the hemlock apart from all other trees. Spruce trees are distinguished by the stubs of old needles which roughen the twigs, giving them a gnarled appearance. Western redcedar branches curve downward; sweeping out at their tips, they give the cedar a droopy, tired appearance.

Some branches seem self-willed, reaching out maliciously. Slide alder owes its shrublike shape to the horizontal sweep of its ground-hugging branches, appendages that may deftly tangle the boots of an unwary hiker. Unwittingly adapted for sloughing off the avalanches of its common habitat, it would be a stretch of truth, although an interesting thought, to attribute to this modest tree a will for aggravation!

During the winter months, the branches of deciduous species serve as a constant reminder of the harsh environment. Against the ashen skies, their stark angular shapes seem to evoke distant memories of a softer form. Vine maple provides contrast to this somber scene; close inspection reveals bright green branches and trunks set against deep red twigs. Bank-side willows, with their whip like yellowish branches, give a warm contrast to the frigid stream banks lined with snow.

With their variable geometry, branches create the droop of a cedar, the long sweep of a Sitka spruce, the stiffness of a Subalpine fir. Designed for a vital purpose, it is the branches that must withstand and persevere through the vagaries of winter weather. Covered with needles or leaves, or bare and vulnerable, the stark limbs support the forest trees.

Needles

The luxurious needles mask a somber beauty.

Within the pigmented cells of the coniferous needle the sun's energy is converted to food through the process of photosynthesis, a chemical transformation that creates organic compounds from atmospheric carbon dioxide and water. These compounds are the food for the individual tree.

Sharp-tipped needles of Engelmann spruce

Symmetrically arranged, coniferous needles may grow as individuals from the twig, as in the case of the true firs, or in bunches, as in the case of the pines. The needles may be flattened or round, blunt or sharp-tipped, short or long. When mature, their predominately green color ranges from bluish to yellow, and they are often marked with white stripes (the stomata) on the top, bottom, or both. Their length ranges from less than half-an-inch, such as in the case of Western hemlock to the long eight-inch needles of Ponderosa pine. Needle density is equally variable. These characteristics not only depend on the family or genus to which the species belongs, but are indicative of the type of habitat that it occupies.

Alaska cedar and Western redcedar bear scale-like needles which hang in feathery fronds from drooping branches. The pine genus is characterized by needles borne in bunches of two, three, or five. The two softwood pine species bear five needles per bundle; these include the remote Whitebark pine, with its yellow-green tufts, and the Western white pine, with its long, soft blue-green needles. Lodgepole pine bears needles in bunches of two, while Ponderosa pine carries needles in bunches of three.

Grand fir needles

Needles are typically slender, with a length much greater than width. They may lie in a single plane (as in the case of Grand fir) or be arranged in a three-dimensional configuration. Some conifers spread their needles out in flat arrays (Pacific yew, Grand fir), while others are arranged in right-angled or double rows (Subalpine fir, Noble fir). The sharp needles borne by the two spruce species probably provide saplings with protection against browsing by herbivores.

The combination of habitat and energy requirements are important factors in the variability of needles. Needles laid out in flat arrays expose a maximum surface area to the sun, enabling them to capture the most radiation during midday hours. Some sun-loving species, such as Douglas-fir, arrange their needles symmetrically around the twigs, taking advantage of all angles. Understory species may compensate for the reduced light with needles of variable lengths spread in many directions from the underlying twig. This strategy is employed by the shade tolerant Western hemlock; its short and variable needles have the effect of increasing the needle density. Other species, such as Pacific yew, survive in their shady habitat with sparsely configured needles, but even this species will assume a dense bush form in more open environments.

The idea that needles are arranged so as to present maximum surface area to the sun falls short of explaining the same mechanism present in trees of sunny spaces. Although the needles are less densely packed than those of Western hemlock, Douglas-fir, a species which thrives in newly cleared areas and requires direct sunlight, employs the radiating method. Two firs of open and mixed forest, Noble fir and Subalpine fir, bear needles which are curved upwards and forward on the twig. The shade tolerant Pacific silver fir combines the flattened and radiating concepts, lining its twigs with flat arrays and covering the top with shorter needles that point forward along the branch. The needles of pine species such as Ponderosa pine and Western white pine are less densely packed; this may in part be the result of the quick growth rate achieved by these species, which results in an open configuration.

Although individual needles continuously die and fall, all of the coniferous species at the Park are evergreen, meaning that they retain their needles throughout the year. Within them, the sun's energy is transformed into the organic molecules that are life's building blocks. Needles permit the passage of air molecules and play a vital role in water regulation. When needles are attacked by insect or disease, the whole tree is threatened with reduced growth or death. Healthy needles are both beautiful and functional. Without the needles the tree dies; with them, the conifer branches are collectively transformed into the largest trees on earth.

Leaves

Temperatures plummet as the wind chills the air and snow blankets the ground. With their buds suspended on stark twigs the deciduous trees seem to abandon all hope, while the conifers struggle for sustenance, their needles exposed to the frosty air.

Winter looms as a hard reality to the trees, and as if unwilling to engage the cold and wet, even Spring seems a reluctant visitor at the Park. By April tentative changes are evident in the lower elevation tree species, as along the streams and wet hillsides the opening of willow buds and the faint green of tall Black cottonwoods tints the air. The alders seem to have given up; holding back their leaves, the pinkish hue of the dangling male part (the catkins) are their solitary contribution to spring color. In the higher elevations, winter tenaciously grips the land. Only the gray stems of Slide alder betray the presence of a deciduous species.

Red alder pollen cones

Beneath the sun's stronger rays, leaves burst open in response, as the brief spring yields to a brilliant summer. An abbreviated season of vigorous activity begins. A great deal must be accomplished during a very short time.

A bud slowly opens to a five-pointed star. Hanging indolently from long stems, the tips are soon transformed into the deeply-cut

Vine maple leaves

lobes on 6- to 12- inch leaves that give Bigleaf maple its name. Bright green in summer, they turn yellow in autumn, and, falling, contribute humus to the forest floor. Similar in shape but less than half the size, the related Vine maple bears seven to nine-lobed leaves, creating a grassy-green star. Growing as an understory shrub or as a near monoculture in avalanche chutes, this species makes its most important contribution in autumn. Then the shrubby maple surpasses all others with a spectacular display of brilliant reds and oranges.

The maples are not the only show stoppers in the fall; Black cottonwoods transform their riverside habitat with domes of gold. The cottonwood's leaves are heart-shaped, up to six inches long; during the growing season, they are colored a deep green with light gray on the underside.

Black cottonwood leaf

The leaves of Slide alder leaves are doubly toothed, with rusty hairs underneath, inconspicuous green in the summer, and dull in the fall. However, beauty has little to do with ecological function, for it is the alders that transform nitrogen in their roots and leaves, an element necessary for all plant growth.

As with coniferous needles, photosynthesis takes place in leaves. When the leaves fall, therefore, photosynthetic activity ceases, and the deciduous tree is essentially dormant until spring. Thus, since light energy is at a minimum in the winter, deciduous trees represent a conservative approach. With little expenditure of energy for maintenance and loss through the leaves, the trees save resources for the time when the energy supply is greater.

With the exception of Slide alder, which typically grows on steep slopes above 3,000 feet, most of the deciduous species at Mount Rainier are restricted to the lower elevations. The leaves of these species do not represent a large proportion of the photosynthetic mass in the forests, and it may be easy to miss them altogether. Along the rivers, under the shade of conifers, in disturbed areas, and on steep slopes their light, rustling leaves betray their presence. At higher elevations, the bright leaves are rare, for it is here that the conifers rule.

Female Cones

The female cones are a distinctive feature of the pines, a class of trees which includes the majority of the conifers at the Park. All pines are gymnosperms, plants which protect their developing seeds with cone scales only, lacking the ovary of the more recently evolved angiosperms. The success of the ancient gymnosperms is easily observed at Mount Rainier where they dominate the forests from the lush lowlands to the upper limits of tree growth near the edge of the glaciers.

Noble fir cones

The female cones are as variable as the trees themselves. Measuring 4-to-7 inches in length, the barrel-shaped cones of Noble fir stand erect near the top of the tree, stiff forms that are an easy field mark for this tree. This is particularly useful, since up to two thirds of the tree's height may be free of branches, making even the lowest limbs difficult to see. Borne within a protective covering, buoyant, long-winged triangular seeds mature in a single season. In the fall, the cones disintegrate and, scattering their scales with "bracts" (the modified leaf-like structure that attaches to the cone axis) still attached, they are reduced to spikes, pencil-like remnants of the stout cones. Noble fir seeds are viable for a year, and thus successful germination is limited to the spring following dispersal. Intolerant of shade, the slow growing seedlings demand abundant moisture. These stringent requirements make them sensitive to both location and climate.

All of the true firs produce female cones annually which disintegrate in place near the tops of the trees. Sometimes the fir cones are cut free by rodents, and may be found on the ground or in a cache. Other species, such as Douglas-fir, Western hemlock, and Mountain hemlock, bear their cones throughout the length of the tree and drop them intact after seed dispersal. The two spruce species (Sitka and Engelmann) are distinguished in the spring by the abundance of purple cones in the upper third of the trees. The two-to-four inch Sitka cones are typically longer than the Engelmann; in both species the cones mature to a light tan color with small, toothed scales. They are somewhat similar to Douglas-fir cones, but the latter have distinctive bracts attached to each scale, and are larger. It may be difficult to distinguish the cones of Sitka from Engelmann spruce; typically, the range of the two species is separated, with Sitka spruce confined to low altitude at the Park, while Engelmann grows at middle to high elevations on the north and east sides of the mountain.

Engelmann spruce cones

Taken together, Alaska cedar and Western redcedar grow throughout the entire forest altitudinal range. They commingle in a narrow band, typically between 3,000 and 3,500 feet of elevation. Both species bear tiny nutlike cones in great numbers. The Western redcedar cones mature in a single season, while the cones of Alaska cedar usually require two years. Thus, Alaska cedar trees bear both first year green and second year brown cones.

Western white pine cones

Members of the *Pinus* genus bear cones which typify the image of a conifer. Requiring two years to mature, the cones are transformed from soft to hard, bristly forms, ranging in size from half an inch to more than six inches. The ridged, thick scales, which vary from gray to tan with age, give the cones their distinctive appearance. With the exception of the Whitebark pine, the cones do not disintegrate, but fall intact to the forest floor. Lodgepole pine cones may remain on the tree for many years; thus the observable cone crop is of varying age. Western white pine bears the longest cones of any

Northwest pine, but only a few are produced each year, and seeds vary greatly in viability. These six-to-eight inch cones hang near the top of the trees.

Coniferous species often reveal their presence with a scattering of cones on the forest floor. Some species, such as the hemlocks, Douglas-fir, and Lodgepole pine, provide easy access to their cones by bearing them on the lower branches. The cone of each species is distinctive--Douglas-fir cones with their bracts, Western hemlock and Western redcedar by their small size, and the true firs by the erect cones at the tops of the trees. Their function is common, however, for all cones carry the seeds necessary to survival of the next generation.

Male Cones

Late in the summer, showy female cones stand erect near the tops of the true firs, or hang pendant-like from the pines, spruces, and Douglas-firs. By comparison, the diminutive male cones are no longer visible, having fulfilled their function much earlier in the season and essentially disappeared. Even during the spring, when the male cones mature, their small size may make them go unobserved.

Male cone — Subalpine fir

The male cones of the true firs develop beneath the cover of the branches, close to the middle of the tree. Subalpine fir, a species which withstands the harshest conditions, produces a bluish-colored male cone barely half-an-inch long. Western hemlock male cones are even smaller; the quarter-inch long structure seems commensurate with the female cone, which reaches at most an inch in length. Each spring tiny cones barely a tenth of an inch long hang from the branch tips of Western redcedar, one of the largest and longest lived conifers; Alaska cedar male cones are only twice this size.

The disparity in size between male and female cones is explained by their different functions. Throughout the growing season, the female cones carry the developing seeds, and, to one degree or another, these seeds must be self-sufficient upon dispersal. Success depends on several factors, including the amount of food stored within the seed. This resource increases seed size, necessitating more space within the female cone. Since reproduction demands resources, the production of the tiny pollen grains within the male cone requires less space, and thus energy is not expended for that purpose. There is little need for size or elaboration. Even the smallest of female cones is much larger than the male counterpart.

The antiquity and range of the gymnosperms serves as testimony to the success of the reproductive strategy. During the spring, when clouds of pollen fill the air, the abundance in this approach is most obvious. Out of sight between the twigs and branches, the male cones make their presence known.

Flowers

It is early in the morning--you know this because the roof of the tent is barely visible in the dim light, you are cold, and it seems like only minutes since you dropped exhausted into your sleeping bag. You shift, hoping to avoid the dewy drops on the tent walls, and moan in reaction to stiffened muscles. Painful memories of yesterday's climb to the ridge on which you are perched remind you of who and where you are.

A hiker shivers in the early light and wonders if summer was missed in the oblivion of an exhausted sleep. In agonizing slowness, the sun inches above the rocky ridge, but, then, as if to reward, a flood of rays pierces the cold and warms the hands. Bordered with dark trees, a meadow is transformed into a kaleidoscope of yellows, purples, oranges, and pinks. The Earth is no longer silent, for as the rising sun suffuses the air with heat, the insects take flight, and soon the atmosphere is filled with their buzzing. Attracted by color and smell, the pollinators start their daily work in mountain meadows. Unveiled in the morning light the meadow glows, and the hiker remembers the reason for coming.

Bitter cherry

Mountain flowers are not available for viewing on demand. Perhaps some of their appeal and beauty to the human observer lies in that temporal quality. Their production, however, is a serious business, one which demands resources that could be expended for other purposes. With a purpose bound to its visual or olfactory appeal, the flower is a complex structure upon which the species stakes its continuance.

Technically, the colorful flowers are the reproductive organs of angiosperms, and probably co-evolved with insects as a new agent for pollination. The seed develops within the flower and disperses. With its mission complete, the flower dies.

Flowers may be arranged singly or in groups, include both female and male parts within the same or separate structures, or place the two on different plants altogether. Conceptually, flowers perform the same function as the coniferous male and female cone counterparts, with a few special adaptations.

While much smaller by comparison to the meadow flowers, the angiosperm tree species at Mount Rainier bear flowers which fill the same function. In some species male and female flowers are borne on separate structures. For example, Bigleaf maple mixes the male and female in a long cluster of greenish yellow, scented flowers. The alders, on the other hand, have separate male and female flowers named "catkins." The male part is pendant , while the female catkin is borne upright.

Willows, too, bear flowers; these are the familiar "pussy willows" that are collected in the spring for bouquets. The female catkins are composed of tiny flowers that form the felt-like rounded head. There are male counterparts, too; these resemble the female with the exception of the yellow stamens that protrude from the soft flowers.

Willow catkins

The angiosperm flower is the counterpart to the gymnosperm cone. Everyone knows, however, that a flower looks really nothing like a cone. Immature cones, the first step towards production of a seed, are barely noticeable, but the flower, possessed of the same function, is attention stealing, and sometimes spectacular. Bright petals, contrasting centers, soft and sharp geometric lines, bees buzzing about--these are what come to mind when we think of flowers.

We write poetry about angiosperm flowers, we climb mountains to see them, we make a business of selling them. A recent invention, they color our world. Their evolution inextricably linked to opportunity and chance, their beauty to the human eye has little to do with natural laws. It is the ephemeral flowers that are assigned the vital task of reproduction.

Forest Variation

Elevation

The Lowland Forest

Aptly named, "Paradise" is the highest point accessible by car. From the parking lot the view is spectacular, but from the road below the mountain is obscured by a backdrop of forest and slope. The trees crowd the shoulders of the road, shrinking the sky to a sliver, shortening the horizon, defining the world.

Giant trees stretch skyward, through the buttresses a silver river shines, and the ancient forest hugs the road. The feeling is difficult to ignore--something is different here. It is calm, quiet, dense. Numberless years slip away as the senses pull the consciousness backwards, straining for timelessness in a clock-mechanized age. There is a feeling of settled seniority, as if all the forest members had attended community meetings for years, worked out the details, and assigned the niches. An invitation to enter comes with a warning--it is easy to get lost where diversity is mistaken as sameness to the untrained mind. Trails disappear within yards, swallowed by the immensity. Stop the car, step out, walk a few feet--and the road is gone.

This is lowland forest, a habitat of potentially enormous conifers, and nearly as impressive leaf-bearing trees. Alongside tempestuous glacial rivers, pioneering Red alders seem to ignore the threat, crowding the sandy flats and hovering over the water. With their droopy limbs hanging down tapered trunks to massive buttressed bases, Western redcedar thrive in wet areas, sometimes forming nearly pure stands. Swaying Western hemlock, thick-barked Douglas-fir, and an occasional Western white pine are at home here, thriving until catastrophe or old age overtakes them.

Lowland forest

Time seems to stand still in this settled land, but the impression of agelessness barely masks an undercurrent of change. On calm days, hikes are quiet times, but a visit during a storm is an entirely different experience. Assaulted by wind and rain, old and young trees alike fall, opening the canopy, and yielding their place to pioneers. Events like these provide the ripples of change that play such an important role in forest dynamics.

The Trees of Mount Rainier

The beauty, the immensity, and the density that characterize the lowland forest are well represented at Mount Rainier. Both major highways pass through old lowland forests, while the Carbon River road gives access to an ancient community, composed of Douglas-fir seven- to ten- feet in diameter! Here these forests grow near the upper end of their altitudinal range. Soon the intermingling of giant conifers and immense cottonwoods yields to a colder and snowier land where trees still grow, but the content and character of the forest change.

The Middle Forest

Awakened by the sun, the land throbs with subtle vitality, obliterating signs of the forgotten winter.

Leaving the lowlands, the trees quickly diminish in size, as new conifers adapted to this colder habitat replace the lowland giants. Gone is the lush forest with its immense cedars; gone, too, are the broad deciduous trees and the oversized river valleys. Cut by steep valleys, glacial moraines, and rampant streams, the landscape grows steeper and cooler. Winter snows linger into late spring. The forests intermingle with an occasional meadow, and new deciduous species thrive in steep chutes.

At Mount Rainier, the transition from the low to middle regime occurs at approximately 2,500 to 3,000 feet, while the subalpine forest begins at about 4,800 feet of elevation. For the hiker, the shift is often gradual, with both the width and elevation variable from one site to another. However, from the vantage point of a car traversing a steep Park highway, the transition is quite abrupt. Reduced in size, the new species compose communities of varying density, sometimes interrupted by diminished meadows, or steep slopes. Tangles of undergrowth, too, are composed of smaller logs and branches, often bleached white in the opening landscape. Snows linger here, sometimes in deep drifts, or as a mosaic of patches. All are part of a changing environment.

Pacific silver fir

The most common species at the Park, Pacific silver fir, dominates the middle altitude forest. Other coniferous species include Western white pine, Noble fir, Lodgepole pine, and Mountain hemlock. Smaller versions of the large lowland conifers, Western hemlock and Douglas-fir, encroach in the lower part of middle zone. Eventually even they disappear, replaced relatives more adapted to the colder environment. Bigleaf maple and Black cottonwood yield to a mixture of deciduous and coniferous species. Vine maple grows on rocky slopes, occasionally in the company of a remnant Bigleaf maple. Red alder is confined to low altitude stream banks and other wet habitats, but its relative, Slide alder, grows in similar sites and on the steep avalanche chutes more common in the middle altitude forest.

Slide alder community

Winter temperatures below freezing are common in the middle altitude zone, with occasional storms depressing the temperatures even further. In these conditions, the heavy winter snow pack is often a beneficial phenomenon. Covering the lower trunks of mature trees and burying seedlings, the snow provides a blanketing protection against the winter weather. Above the snowpack, saplings are subjected to more widely fluctuating temperatures and conditions than within, where a nearly constant temperature may spell the difference between life and death.

The snowpack may impact the range of species. Pacific silver fir is reduced in stature with increasing elevation, and is restricted from the highest forests altogether, a change that may, in part, be a function of snowpack. Conversely, tree species more commonly found in the subalpine forest may be not compete as well in this regime.

Spring is a time of transformation in the middle altitude forest. The increasing solar radiation slowly dispenses with the snowpack, and exposes the young trees, while providing a warm substrate for the germination of the previous season's seeds. This is a time when the middle altitude forest is brightened by pale flowers and soft understory greens. With the increasing solar radiation, the production of new cones begins. It is a season infused with energy.

Except for isolated patches, during the summer the snowpack disappears from the middle altitude forest. With greatly reduced rainfall, the species must endure a dry season. After a warm and wet spring of vigorous growth, photosynthesis often decreases significantly. This is a conservative season for trees and shrubs, in which the reduced moisture may threaten survival, particularly for the seedlings.

Lingering snowpack

During the ice ages, middle altitude species moved into lower elevations as the glaciers at Mount Rainier extended downward. Today, those species push upwards against receded glacial snouts, spatially dominating the forests at the mountain. Here, the white lines of Pacific silver fir needles cast a bluish tinge to the landscape, while Vine maple emblazons the rocky chutes with a fall color show, and Slide alder pioneers glacial borders. This is a land of opportunity and chance, a zone where in select locations, over half of all the coniferous species at the Park grow in close proximity.

The Subalpine Forest

From the buildings with their pitched roofs, to tree spires that poke the sky, Mount Rainier defines the subalpine environment. This is timberline, an area of diminishing forest and lush meadows, where within a stone's throw from paved islands, the plants yield to the power of grinding ice.

On the eastern slopes of the Cascades drought tolerant Ponderosa pines create a spacious forest between the desert lowlands and the high peaks. As the elevation increases, a wetter and cooler climate favors other species, and the pines lose their dominant status, eventually disappearing altogether. Crossing the Cascades to the west, the climate grows wetter still, maintained by moisture-laden clouds from the Pacific Ocean. Situated west of the crest, Mount Rainier is a landscape seemingly ill-suited for pines. And yet, surprisingly, pines grow here, including the Ponderosa.

Ponderosa pine

There aren't many Ponderosa pine trees at the Park. Their presence on the eastern side of the mountain may be due to dispersal by animals from the east side of the crest. It might also be attributed to the generally drier conditions here. Partially shielded by Mount Rainier from weather fronts carried on southerly winds, this region receives less snow than the southern side. The earlier melting of the reduced snow pack increases the potential growing season, enabling species to extend their altitudinal range. Here the related Whitebark pine does very well in the highest regimes.

Near Sunrise, Whitebark pine pushes the boundaries of tree habitation to elevations well above 6,000 feet. Harsh conditions at this altitude sometimes force this species to adopt a shrublike form, but in more protected locations, Whitebark pine spreads its multiple branches into a wide, bushy crown. Drought tolerant and adapted to thin soils, this pine grows in areas of recent volcanic pumice deposits.

Whitebark pine

Engelmann spruce also grows exclusively at high altitudes on the east side of the Park. Surviving on a rock-strewn slope above 7,000 feet in elevation, perhaps the loftiest of any tree at Mount Rainier, the shrubby spruce bears little resemblance to others a thousand feet below. Like other trees in this extreme environment, the age of the spruce is difficult to determine, as growth may be exceedingly slow, but the tree hangs on, healthy in appearance in spite of its reduced stature.

Third Burroughs Mountain — a site for Engelmann spruce

Other subalpine species are more cosmopolitan in their distribution. Alaska cedar forms dense communities on steep slopes and in the fog and mist zone of ridge tops throughout the subalpine region. Mountain hemlock finds sufficient moisture in select locations, and Subalpine fir grows throughout the region, too, sometimes in forested stands with Mountain hemlock, but often as tree "islands" in the meadows, and, in more extreme environments, as wind-sculpted shrubs.

On Mount Rainier's broad south flank a paved road ends abruptly in a concrete field criss-crossed by white lines and surrounded by vintage national park buildings. This is Paradise, at 5400 feet the subalpine site most frequented by human visitors. From Paradise the mountain is so close as to give the impression that one could climb it directly, with little intervening territory to cross. And, in fact, this is true, except that climbing involves crossing immense glaciers. Without climbing the mountain, however, enthusiasts and casual onlookers alike may observe the effects on a land under siege. In this seeming contradictory environment, where flowers sprout near a glacier's edge, the trees extend their domain to the highest possible limits, appearing as dark dots along ridges above moving glaciers. These are the trees that define the subalpine forest at its greatest extent.

In late summer, when the last sliver of snow has melted beneath the sun, the meadows are transformed into a multi-colored panorama of green grasses and ephemeral flowers. Interspersed into this display are clumps and lines of dark-needled Subalpine fir and Mountain hemlock. These trees are often much smaller than members of the same species in habitats only a few hundred feet below. With increasing altitude, the firs and hemlocks continue to shrink, often only surviving as shrubs. On a ridged backbone, they grow on terrain over a thousand feet higher than the parking lot. Free of snow for only a couple of months each year, the growing season is exceedingly short. The distribution of these pioneers is not necessarily random, however. The ridge tops are typically snow free earlier in the spring than the surrounding slopes. This phenomenon may be attributed to both sliding or removal of snows by wind, effectively reducing the snowpack.

Ridge-top trees

The Subalpine fir trees are distinguishable as narrow spires, growing in close proximity. Mountain hemlock is common but not as abundant, while Pacific silver fir adopts a similar form (although typically taller) as Subalpine fir. Wind and ice strip the windward side of many trees. Strange black molds besmirch the needles but typically do not kill the trees. Cloned saplings grow down slope from the parent trees, where they survive small avalanches in a continual invasion of the meadows.

Forest Variation

Mountain hemlock

Traveling around the mountain, the highest forests vary in species membership and extent. This variation results from many factors. Disturbances such as fire and avalanche create openings. Recently deglaciated terrain permits pioneers such as Lodgepole pine. Regional ice age dynamics may influence other species, such as Englemann spruce. And eastern species, like the Ponderosa pine, grow in small numbers in meadows and forests. Thus, a landscape that may seem harsh nevertheless is a place of variability and opportunity.

Habitat

Ridges

A geographical line between slopes of different aspect, ridges often create environments sufficiently diverse to support unique plant communities. For example, near timberline ridges may divide regions of forested and non-forested terrain. Both aspect and steepness influence placement on these slopes.

Slopes reveal a geological story linked to climate and topography. During the ice ages, massive glaciers blanketed the landscape, engulfing Mount Rainier and extending into the surrounding lowlands. The ice cover was not of uniform thickness, however, and, as the climate warmed, the glacial melting and the subsequent withdrawal up Mount Rainier's slopes varied. Glacial remnants today reflect this history, and the mountain itself is pock-marked with glacial patches on high, sun-starved, north-facing slopes.

Receding glacier — Observation Rock

While slopes with a southern aspect often support continuous forest, in addition to the cold, the steepest north-facing regions are slide prone. Periodic avalanches may sweep the terrain clean of all but the most resistant species. In addition, the lingering snow pack on the north slopes may inhibit germination while only a few a sunnier and gentler terrain supports life.

At Paradise, the undulating topography is populated with a patchwork forest of Subalpine fir and Mountain hemlock. On ridge tops and the adjacent steep slopes dense forests often end abruptly where the inclination of the slope decreases and grades into a shallow valley. This pattern seems inconsistent, for it might be expected that the less gentler slopes with their deeper soils would provide a more hospitable environment than the steep, rocky terrain. However, steep slopes may readily slough the winter snows into the swales at their base, effectively decreasing the snowpack on these slopes. The valleys, as recipients of the sliding snow, remain covered much longer, shortening the time available for seedling establishment during spring and summer.

Denuded slope below ridge

Successful pioneering in such marginal habitats may depend on the time span between the final disappearance of one winter's snowpack and the onset of another. A succession of dry years may promote germination and establishment in areas otherwise inhospitable. If the trees do survive during this time, they will be in a position to withstand a return to harsher conditions.

Not all north facing slopes are treeless. If the slope is sufficiently gentle, the lingering snow pack in these areas may promote species excluded from the southern aspect. The ridge line itself is an important habitat for tree species adapted to the subalpine environment. Stunted by the wind, limited by shallow soils and a short growing season, they grow to the glacial edge.

Wetlands

In the Northwest, most wetlands are found at low elevations, where poor drainage or insufficient outlets inhibit runoff or absorption. Some wetlands may become dry in the summer, while fall and winter rains engorge these habitats and often increase their size, creating new areas of standing water. This phenomenon is particularly observable in river deltas where seasonal overflow is frequent, and enlarged wetlands persist for months. Evidence of the inundation is recorded in the gray soils and unique plants characteristic of these environments.

Since wetlands are most common in lowland depressions, the steep, glaciated terrain at Mount Rainier would seem to preclude their formation.

However, although not common in the Park, well-developed wetland habitats are present. Topography conducive to their formation includes gentle slopes and flat benches where drainage is poor or a surplus of water if present in springs and streams. In the subalpine environment, these areas are often located in north-facing regions where the winter snow pack may linger, or where snowmelt accumulates along swift glacial streams.

In lower elevation forests, marshy areas are home to stands of species tolerant of the inundated conditions. These species may be indicative of the wetland's demise, as soil buildup promoted by the trees, and the invasion of new species transform the wetland to a dry depression. Where the process is reversed, recently formed wetlands are recognizable by the presence of stark and dying trees--drowned ghosts of a drier landscape.

Middle elevation wetland

Wetlands are important in providing habitat for species which might otherwise be excluded. On the north side of Mount Rainier, Engelmann spruce grows in moist, high elevation habitats, and Alaska cedar occasionally attains a rare dominance in wet depressions. Robust Western redcedar also thrives in low elevation wetland habitats.

A disturbance species, Lodgepole pine forms densely packed communities in regions that are frequently swept by fire. At Mount Rainier, however, Lodgepole pine must compete with fast growing species such as Douglas-fir. Both are pioneers, quick to invade disturbed areas, requiring ample sunlight, and adapted to a range of habitats. Unable to match the growth rate of Douglas-fir, Lodgepole pine is typically delegated to a minor pioneering role, often in areas that are not only recently disturbed but inhospitable for other reasons. In these environments, Lodgepole pine may be the only tree species present. An example is in the Longmire meadow, on the Park's south side. Here, with their de-

Lodgepole pine at Longmire

posits of toxic chemicals perennial hot springs create an inhospitable environment. Encircled by a forest comprised of several tree species, the wide meadow is populated with tolerant grasses, shrubs, and a few Lodgepole pine trees. This species grows to the edge of the wettest meadows as well as on the hillside. Recent fires may have contributed to the success of this species, while their distribution may indicate a continuing selection for areas of reduced chemical emissions or soil accumulation.

Sitka spruce

A species of the wet coastal lands, where it attains immense size, Sitka spruce also grows at elevations greater than 2,000 feet at Mount Rainier. These lowland wetland habitats include the lower reaches of the Carbon River, where Sitka spruce grows in an ancient forest. It does so vigorously, for although the individuals here are not of great size, they are robust and strikingly beautiful. Sitka spruce also thrives in the frequently flooded areas along the White River near the Park boundary, as well as in similar habitats close to the entrance along the Nisqually River. Saplings in all of these locations indicate a continuing presence for this lowland species. Alongside Highway 410 Sitka spruce also grows near Engelmann spruce, the smaller species of middle to alpine habitats.

Subalpine wetland

Although relatively rare, wetlands are the product of many factors. Climate, glacial and volcanic history, seasonal and annual fluctuations—all of these make their contribution. The wet meadows, small ponds, and wandering streams provide interesting contrast to the dense, middle elevation forests and the open slopes of the subalpine zone. Walking the fringes of these marshy regions provides the opportunity to view these unique habitats.

Forest Variation

Rivers

Meandering is a luxury reserved for ancient rivers that flow through flattened valleys of their own making. With a gentle slide into old age, these are the water bodies that form arching oxbows, where rapids are rare and placidity is the rule. They are not the rivers of Mount Rainier.

Hike to the high swinging bridge over Tahoma Creek and peer down at a "creek" gushing from the gray jaws of a glacier by the same name. It is very noisy here above a stream that poses a threat to anything within reach, and to much beyond. A menace to the struggling forests lining its banks, near the glacial snout, the river gnaws at the debris, cutting a deep chasm. Perhaps slightly less voracious but no less active, the Carbon River throbs through a wide valley that reflects the ancient glaciology of the region.

These are serious, young rivers, seemingly intent on making their own history. They are brown and gray, and riotously tumultuous. Their waters tumble the mountain's rocks, continually rearranging their beds. They present no invitation for a swim in a quiet pool, for there are no pools.

Forest alongside glacial river

Rivers that originate from active glaciers are colored by the load of sediment entombed by the glacier and released during melting. Typically milky in appearance, most of the major rivers at the Park are of this type. By contrast, snow melt and ground water are the source of many of the smaller streams; these flow clear and clean over their rocky beds.

Another distinguishing feature of the glacial rivers and streams is their swift flow. This characteristic gives the narrower glacial streams the appearance of a much larger river. While pools are more common to the non-glacial streams, the swift waters often pass through precipitous terrain, forming waterfalls and cascades. In this way, their course reflects the geological complexity of the Park.

Wide river beds traversed by slender glacial rivers provide a glimpse into the glacier's history. Here the layered river banks are cut open like an ancient book, revealing their geological story. With a tendency to wander about its wide bed, the river undercuts the bank, leaving damaged or fallen trees at the river's edge. Occasionally overflowing its boundaries, the river drowns trees in the accumulated sediment. Stark limbs serve as reminders of the deluge.

Borne in a cool era, and fed by the wet climate of today, great glacial rivers sculpt much of the topography at Mount Rainier. Mesmerizing and terrifying with their power, these rivers seem only remotely related to the calmer flows of other regions. Along their banks, trees grow tall, only to be swept away. Others thrive in the sand and rock, away from the water's edge—for the time being.

Aspect -- North, South, East, and West sides of the Mountain

From the corner closest to the sea the Nisqually River departs from the Park through a wide valley of its own making. It leaves in a much quieter mood than where it left its glacial source over 3,000 feet above.

Most tree species are cosmopolitan; that is, they are found throughout the Park where the specific conditions to which they are adapted are present. For example, Douglas-fir grows in all the major river valleys, as old individuals in established forests, and as younger pioneers in more recently disturbed sites. Pacific silver fir likewise grows in middle altitudes on all sides of the mountain. However, there are regional differences for some species. Grand fir provides a particularly interesting example.

Old Grand fir trunk

Grand fir grows in several major river drainages on the north, west, and south sides of the Park, although it is absent from the southeast valleys. At Sunshine Point Campground, near the Nisqually entrance, Grand fir is an important forest component, and occurs in small numbers near the river as far upstream as Longmire. The southwestern location of the Nisqually valley may explain the presence of Grand fir, as this species possibly pioneered from the Puget Sound lowlands. Prior to extensive logging, Grand fir was undoubtedly more common along the river outside the Park; old individuals are found in the town of Elbe downstream.

The common lowland deciduous species populate the Nisqually, including Red alder and Black cottonwood. With increasing elevation Red alder is replaced by Slide alder, and the cottonwood and Bigleaf maple disappear altogether. At these higher elevations, too, Western white pine is found in disturbed sites, such as the Kautz Creek valley, where an enormous mudflow denuded the land more than 50 years ago. Vine maple becomes more common in avalanche chutes, and the willows inhabit wet slides.

Above Longmire, a recent burn opened a site for several species, including a large community of Lodgepole pine, and a mixture of low and middle altitude forest inhabitants. One of the most interesting forests along the river is found near Cougar Rock Campground; here an old growth forest is distinguished by origins that date to the receding of the Nisqually Glacier some 1,100 years ago. This, then, is truly a pioneer forest, growing on previously uninhabited terrain. The size of the old Douglas-fir trees here does not approach that of similarly aged trees found elsewhere; this may in part be attributable to the elevation, but equally may depend on the forest history of the site. Pacific silver fir is the regenerating species in the understory, but the old silvers, too, seem diminished in stature.

Western white pine

Near the Nisqually overlook, at an elevation of 4,000 feet, the forest is dominated by middle altitude species, although not far downstream an old Douglas-fir crowds the road. Species more typical at higher elevations, such as Subalpine fir, make this a truly potpourri forest. Noble fir is abundant here, too, and sometimes forms nearly pure stands.

On the Park's southeast boundary, the Ohanapecosh River departs at an elevation just under 1,800 feet, the lowest in the Park. A clear river with a source in a dying glacier, the Ohanapecosh swells from the melding of several substantial streams. The high altitude of these creeks provide habitat along their banks for species such as Subalpine fir, Mountain hemlock, and Alaska cedar. Passing through the ubiquitous Pacific silver fir forests, at lower elevations the Ohanapecosh valley widens and flattens. The forest is dense and rich at these lower reaches of the river, populated with lowland deciduous species and a robust community of middle-aged Douglas-fir and younger Western hemlock. Absent, however, is the Grand fir found in other river valleys.

On the northern side of the mountain another uncommon conifer at the Park, Sitka spruce, grows in wetland areas near the Carbon River and along the North Puyallup River. Exiting the Park at an elevation of 1,900 feet, the Carbon passes through an ancient forest composed of several Douglas-fir trees eight and nine feet in diameter, stout Western hemlock, an occasional

Grand fir, and Sitka spruce. The river has carved a wide bed, through which it cuts channels along sparsely covered sand bars. These raised areas support a mixture of Douglas-fir, Western hemlock, and Western redcedar, Red alder, and Grand fir. Farther downstream a community of large Grand fir indicate the long term presence of Grand fir in the Carbon valley. Grand fir is also established on the more northwesterly aspect in the valley of the North Mowich River. Here the river passes through sandy banks and glacial outwash; like the Nisqually, this valley lies closer to the Puget Sound lowlands, an area of Grand fir establishment in the past.

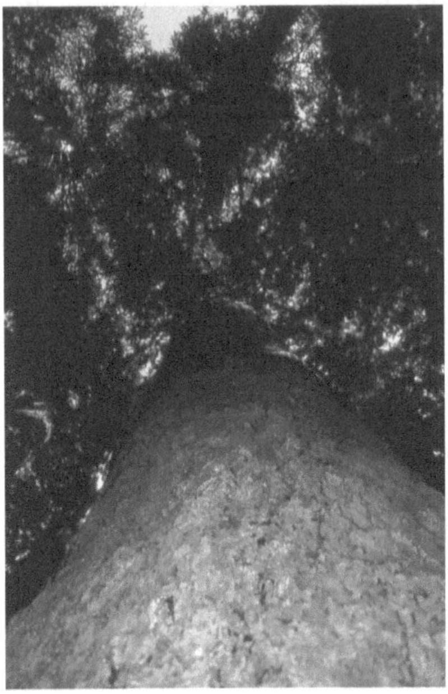

Big Douglas-fir

On the mountain's northeast flanks the White River departs from the massive Emmons Glacier. This valley is a region of lower than average temperatures and precipitation. Engelmann spruce grows at elevations above 4,500 feet in the drainage basin of this river. This spruce is confined to this side of the mountain where its presence may be attributable to the proximity of the abundant spruce forests east of the Cascade crest.

The White, the Carbon, the Nisqually, the Ohanapecosh--these are major rivers threading through wide valleys sculpted by their glacial sources now high above the valley floors. The occasionally striking differences in the forests along these rivers reflect a regional history of disturbance, geological variation, and glacial activity. Related to that history, climatic fluctuations over short and long term scales exerts a strong influence on forest composition. Small regional differences may permit a pioneering role of species that otherwise might be excluded, while long term stability within the species' tolerance promote uniformity. Proximity of the lowlands to the west or the Cascade Mountains to the east influence forest composition, too, and increase regional variance throughout the mountain's perimeter.

Forest Processes

Overview

At an elevation of approximately 2,000 feet, the glacier-carved Carbon River valley is home to a forest at least 500 years old. Ranging up to nine feet in diameter, some of the largest Douglas-firs in the Northwest grow here. Hundreds of years old, the trees are marked with black scars along their red bark, signs of an ancient fiery battle. In time, they will succumb to age or disturbance and die. Evidence of these ongoing deaths is revealed in the immense logs scattered on the forest floor.

Carbon River valley

The thick bark of old Douglas-fir trees provides some protection against fire, but other physical factors, such as wind, sun, rain, and gravity, also influence the structure and function of the trees. The factors are not only negative but may provide opportunity for the species. Douglas-fir seedlings are intolerant of shade; these giants probably germinated in a sunny spot, perhaps one created by fire.

Geological events such as volcanic eruptions enable species adapted to the porosity of volcanic tuff to become established; Lodgepole pine and Whitebark pine often grow on this type of substrate. Periodically swept clean by gravity acting upon snow and mud, avalanche paths are nonetheless suitable habitat for resistant species. In autumn, the profusion of color on these slopes is that of Vine maple leaves, an avalanche survivor. Slide alder, too, grows in these unstable sites, bending rather than breaking beneath the onslaught.

In combination, wind and gravity carry and deposit the winged seeds of Douglas-fir to new locations. Gravity and water may also aid regeneration by means other than seeds. Branches are carried down swift streams, where, grounding on a bank, the branches may take root.

Physical factors may affect trees on time scales measured in thousands and millions of years. The ice ages undoubtedly influenced tree evolution

Avalanche path

in the past. Long term changes in global climate have promoted the emergence of new species, while causing the demise of others. Voluminous rivers and streams, as well as permanent and fluctuating ice fields influence forest composition. Volcanic processes alter local topography and the communities that inhabit the new land. The smallest of fluctuations in any of these processes may have an impact on the trees.

Cellulose, the building block of trees, facilitates size and structure. The efficiency of the food and water transport system also contributes to the great stature attained by many species. Soil, too, is in part a product of the interactions of a host of biological organisms that act upon organic debris.

A host of harmful agents target trees while toxins within tree bark promote longevity, thus enabling many species to reach great size.

Other biological factors promote regeneration. Seeds are important food sources for many birds and mammals; large seeds in particular are important. Fruits are particularly attractive; following consumption, the seeds within may pass unharmed or even enhanced through the animal's digestive track. The success of this method may more than offset the energy required for the fruit's production, and defines a relationship between plant and animal with survival implications to both.

Layering is an important cloning mechanism for some species; in this case an entire community of identical offspring will surround the host. This process is especially evident in subalpine forests where protection is provided within the immediate environment of the parent tree.

Protection within a small community

Tree species may enhance their own reproductive success and that of others. The impoverished soils of avalanche chutes may handicap Slide alder, a common species of such sites. Along with its relative, Red alder,

Slide alder compensates for the drawback by "fixing" nitrogen, an element necessary for life, directly into the rudimentary soils. Where glaciers have receded, the nitrogen contributed by Slide alder to the rocky substrate aids the establishment of species that require the enriched soil.

Hemlocks on old log

Eventually even the most resistant of trees dies, but death itself is not necessarily an ending. Sprouting from old logs, saplings receive nourishment and moisture from their host. Animals utilize decaying snags, and soil building organisms convert the remains. And openings created by a fallen giant are sites of opportunity to species requiring abundant light for growth.

Physical and biological processes affect tree species in many ways. Some may be of little impact to one species while playing a vital role in the establishment of another. Adaptations are not isolated events but are also influenced by other plants, as well as the animals that define the forest community.

From the white tufts of Black cottonwood, to the hidden but forgotten squirrel caches of Noble fir seeds, trees employ various reproductive methods. Girded by the strength of their molecular structure, trees stretch skyward; with flexible limbs, they sway and adapt to the demands of wind and rain. In death, they provide for other forest members. Finely tuned by millions of years of evolutionary history, trees have withstood the test of time.

Biological

Dispersal

In late spring, the green branch tips of Western hemlock are transformed to purple hues beneath a profusion of female cones. By autumn, the cones have matured to a tan color and are ready to shed their seeds. Following dispersal the inch-long cones drop to the lowland forest floor. So small as to go unnoticed, they litter the ground, indicators of a prolific Western hemlock nearby. Young hemlocks may begin cone production at twenty-five years of age, and although the seeds are small (300,000 per pound), a single attached wing can carry the seed a half-mile from the parent tree. Successful germination is most common on fallen trees (nurse logs). Often revealing their presence with a neat row of sprouting hemlocks, these logs provide both nutrients and freedom from competition. Although few of the vast numbers of seeds dispersed each year will germinate, and even fewer of those that do will reach maturity, a robust, multiple-aged stand of Western hemlock in an old lowland forest attests to the success of this method.

Western hemlock female cone

Species other than Western hemlock produce tiny seeds. Western redcedar seeds are even smaller (400,000 per pound), with a tiny wing that carries them only half the distance from the parent tree as the Western hemlock. In spite of the small size, Western redcedar does not produce as many seeds as Western hemlock. The longevity of this species, and its ability to inhabit wet habitats often inhospitable to other species, as well as cloning, are additional factors in the cedar's success. A relative, Alaska cedar, produces a seed nearly four times as great in weight, but does not provide a sizable carrying wing for dispersal. Similar to the Western redcedar, the long-term survival of these disease and insect-resistant trees is aided by a long life span.

While the numbers strategy implies less dependence on the individual cone, the small seeds typical of these prolific species supply less nourishment to the germinating plant, a potential disadvantage where resources are limited.

Female cones —Alaska cedar

The female cones of the true firs grow upright in the highest branches, far above the forest floor. In autumn, the cones disintegrate in place as the large seeds are released. The tree-top location provides the heavy seeds with a potentially wider dispersal range than that available below. Winds are often stronger in the canopy, a phenomenon easily observed on a day when the tops of the trees sway in the breezes while the air near the ground remains still. Wind also promotes the dispersal of pollen from the male flower to the female cones, a necessary step for seed formation.

Animals aid in the dispersal of

Subalpine fir female cone

Western white pine and Whitebark pine seeds, often at a considerable distance, or sometimes stashed for later consumption. The pines produce few seeds in comparison to the other conifers – Western white pine may bear less than fifty cones each year. Therefore, utilizing animals for seed dispersal may be an efficient method, since the tree might not expend as much resource for cone production.

In contrast to the small coniferous seeds, it is hard to miss the piles of fluffy white Black cottonwood plumes which line riverbanks and trails in the summer. Buoyant even in light breezes, the seeds seem to float. As with the related willows, Black cottonwood is typically confined to streamside and other wet habitats, a constricted range that perhaps places added importance to a wide dispersal capability.

Advantages to a reduced dispersal range include the protective environment of the parent tree. Employing animals for the dispersal of large seeds may work well in some habitats but be relatively unimportant in others. In addition, this method may limit options more than the use of physical forces, as plant immobility limits in scope even the most clever dispersal scheme.

Whether produced by the millions or the thousands, only the smallest number of seeds are destined for maturity. Likewise, the rate of forest expansion is limited; this fact is evidenced in the northward migration following the end of the most recent ice age. On a shorter time scale, immediate opportunities may be provided by local disturbances, such as fire or glacial melting.

The constancy of natural factors as well as the short term variance in those forces have ensured a continuing role for a relatively limited number of species. In addition, successful dispersal is dependent upon the continuation of both physical and biological forces. Variation of conditions could result in a shift in forest composition not only in terms of the current species but in the evolutionary strategies that bring reproductive success. What appears to us today as a resounding success story may, in the future, become a historical question.

Germination

With clever schemes, the aging trees cast winged seeds to the winds.

Dispersal strategies are only one step to successful germination. Although dispersed from the tops of the trees, true firs produce only a few heavy seeds, thus limiting the number available for regeneration. Other benefits may be available to a seedling, however, and the small number of seeds may have evolved in reaction to the unique stresses and opportunities of fir habitat. For example, the harsh environment inhabited by Subalpine fir is restrictive to many species, so that prolific seed production may be both unattainable and unnecessary in a region of reduced competition. While only a few seeds may successfully germinate, time and habitat are factors in their favor.

Another true fir – Noble fir – inhabits the middle elevations, where the rapidly growing trees may bear cones in as few as twenty years of age. Many of the seeds are not viable, and their large size (13,500 to a pound) limits their dispersal to the immediate vicinity of the parent tree. But there may be little advantage in germination at a distance, as Noble fir seedlings require ample sunlight for growth, conditions most likely available in the open habitat near the parent tree. Although the reproductive strategy is similar to its more widely established true fir relatives, the exacting requirements of Noble fir, and the size of the seeds, may explain in part the limited distribution of this species.

Cones on top of Subalpine fir

Some species produce seeds that remain viable for many years. Lodgepole pine seeds may germinate following a fire after seasons of encapsulation in cones. Other seeds require cold and moisture prior to germination. Since most coniferous seeds ripen in late summer or fall, germination is often delayed until the following spring. Thus, most endure a cold period and are released from dormancy only when the days grow warm.

Many species have restrictive requirements for the seedling development. Douglas-fir prefers a mineral soil free of organic debris; Western hemlock seedlings often thrive in elevated sites such as those found on fallen logs. Western hemlock may also germinate in small numbers in disturbed areas, growing alongside Douglas-fir and other species which require more direct sunlight. The presence of Western hemlock at these sites may be important for its success as a climax species. Mature trees will produce offspring that are adapted to the shaded conditions of the aging forest, survivors over their less shade tolerant competitors. Pacific yew is one of the few species that germinates well in organic matter, a preference that might reflect its role as an understory tree.

Pacific yew tree

Time of germination varies considerably among the deciduous species. Black cottonwood seeds mature in late spring. On moist substrates, they will sprout immediately. Bigleaf maple seeds ripen in the fall, with germination in winter or early spring; Red alder behaves similarly, although the catkins of this species are often the earliest of the spring flowers.

Seedling development depends on many factors, including moisture, sunlight, competition, substrate, and the presence of animals, insects, or disease-carrying organisms. With risks so high, the percentage of seedlings that survive the first growing season is small. Prolific seed production may testify to the hazards of reproduction, but if a large percentage survived, an aging forest would appear densely packed with young trees of multiple ages. In fact, trees are well-spaced in old forests, and the number of young trees limited. From a forest community standpoint, limitations on germination becomes a strategy for resource allocation and maturation.

Nurse Logs

Amidst what appears as a lackadaisical order in the forest, occasionally a straight line of saplings will grow slightly elevated above the forest floor. On closer examination, the young trees are observed to be rooted on a fallen giant, laid horizontally between living trees, many of them scarred by its fall. In death, nothing has been wasted, as the ancient tree now provides a germination site and source of nutrients to the young trees. In its new role as "nurse log" the old tree fulfills a vital role in the forest community.

There are advantages to growing on a rotting log rather than the forest floor. Through the decay process, the disintegrating structure may provide a steady water supply and reservoir for the seedling. This storage capacity is especially important during the summer drought. Nutrients are released by the breakdown of woody material, and a suitable environment for beneficial fungi is provided. The elevation of the log above the forest floor may make the seedlings less susceptible to smothering from collected debris or competing vegetation. With their fall, nurse logs create holes in the forest canopy, opening a shaft for sunlight, and providing an energy boost to the young trees.

Fallen trees

In many lowland forests, the greatest percentage of successful Western hemlock seedlings germinate on nurse logs. Western redcedar, Sitka spruce, and Grand fir are among other species that utilize nurse logs, sometimes growing next to each other on the same log. The young trees often attain considerable size before the supporting log disappears completely, subsiding with the row of saplings as a grave marker. Extending their roots downwards in a continual search for food and water, these trees support buttressed trunks as mute evidence of their beginnings.

Nurse logs demonstrate that little is wasted in the forest. Through its death and subsequent utilization, the fallen tree proffers evidence of interdependencies and intricate, if unconscious, design.

Disease

There is a wide range in tree longevity. Old age varies by a factor of twenty or more, from 50 to greater than 1,000 years. This span suggests a question -- if trees species have evolved to endure for ten centuries, why do not all survive that long? A partial answer is found in the differences between angiosperms and gymnosperms, but even within those orders, the onset of senescence varies greatly.

Eventually, even the most ancient tree dies, and if the cause is not the result of large-scale disturbances, such as wind or fire, it is attributable to years of persistent attacks by insects, diseases, or both. From the shortest-lived Northwest species (Bitter

Diseased Red alder

cherry, an angiosperm) to the longest (Alaska cedar, a gymnosperm), all tree species have their enemies.

An example is found in the beautiful and versatile true firs. Although other species exude a resinous pitch as an aid to closing wounds, the firs do not produce this substance and thus have little resistance to invasion by fungi. Anything which pierces the thin bark opens a potential hole for opportunistic parasites, insidious infections which, with time, expand their spatial hold. When the bark of Douglas-fir is penetrated, this long-lived species seals the break with sticky resin. In spite of the lack of pitch, one true fir – Noble fir – apparently puts up sufficient resistance to disease to make this species the longest-lived of the true firs, although its 600-year life span still falls short of the 1,000 years of Douglas-fir.

Hole in old Douglas-fir

There are hundreds of fungi named to describe appearance or function. Western hemlock succumbs to a bole-rotting fungi known as yellow ring rot, while the life span of Grand fir may be shortened by Indian paint fungus, a large conk whose presence on the side of the trunk is indicative of internal rot. The longevity of Western redcedar and Alaska cedar can in part be attributed to the toxic chemicals that resist fungal attack. Although most are minor threats, Black cottonwood hosts at least seventy fungal species. Old Bigleaf maples are susceptible to both core and root rots.

An invasion of Dwarf mistletoe imparts a grotesque appearance to many Western hemlock trees; this parasite is a relative of the more popular Christmas mistletoe. Invading weakened trees, the mistletoe creates large clumps on the branches and stunts their horizontal growth; the dark round masses created by the mistletoe can be easily seen. The mistletoe plant expels its seeds from little capsules, scattering them sufficiently to continue the infection amongst healthy trees. Lodgepole pine is especially susceptible to mistletoe, which is capable of sprouting on most conifers; even the resistant Douglas-fir is occasionally a victim.

In the subalpine zone, black molds darken the branches of Subalpine fir

Fungal fruiting bodies on rotting wood

and Mountain hemlock, and although not fatal, an infected tree appears to be in poor health. Whitebark pine, a high elevation species, hosts an exotic disease, blister rust. This disease was imported from Europe on white pine seedlings and has proven deadly to both Western white pine and Whitebark pine. Killing the tree, the movement of infection through a stand of trees can be observed in the presence of many standing pale snags.

Starting early in life, most trees are probably infected with internal rot to one degree or another. Death comes incrementally from the slow growth of the invasive organism, a long term attack so that even the most short-lived coniferous species may live for over two centuries. Many factors contribute to this longevity, including wood strength, bark thickness, and the presence of sealing resins and inhibiting chemicals. Often as much dead as alive, these defenses enable old trees to withstand the assault of disease and decay for many years.

Trees employ a multitude of defenses against the attack of agents, whose own continuance may depend on the host. But no defense can guarantee immortality, and, eventually, even the most ancient of trees succumbs. But even in death they provide opportunities for life in the openings they create.

Physical

Climate

The shoulders of the road whitened, as the windshield wipers labored against the thickened rain. The skies lowered, promising a prolonged storm, and the conifers darkened beneath a powdered dusting. Their growing season now ended, the trees bent beneath the load, their continuance staked on the moisture within the protective white blanket.

Positioned west of the Cascade Mountain's backbone and more than 14000 feet in height, Mount Rainier's is influenced by two major climatic factors. Both seasonal and global, the first owes its origins to the Earth's rotation and inclination towards the sun, as well as the position of the land and sea masses. The rotation creates lofty winds (the jet stream) which dominate the weather patterns with winds that shift seasonally. In the winter, a southerly air flow laden with moisture from its passage over the Pacific ocean deposits its contents as rain or snow. During the summer, the winds shift to the north, and drier air brings a period of drought.

The bulk of Mount Rainier creates the second weather effect. Clouds rise around and above the mountain, relinquishing their moisture, and localized winds create a micro-climate. Although by comparison to other places at similar latitudes the winters are mild and very soggy, storms and cold nevertheless limit the range of moisture-loving species. At higher elevations accumulations of heavy snow break tree limbs and bury young saplings. And the snow often lingers long enough to enforce a brief growing season, thus determining not only which species will flourish, but where, in a geographical and habitat sense, they will grow.

Heavy snow on Subalpine fir

The summer season is also influenced by the mountain. On the east side, the climate is drier and the timberline higher, thanks to the blocking of prevailing westerly winds. The occasional onset of sudden storms and reduced temperatures are a source of stress for the trees. Even in the middle altitude zone, snow may melt slowly in protected locations, inhibiting growth.

Trees killed by fire

Local topography impacts air flow and temperature. An example is found along stream valleys where the drainage of cool air and excess moisture may determine the presence of a species outside its normal range while at the same time limit others. Both the precipitation phase (rain or snow) and the rapidity with which it melts are influenced by local conditions. South-facing slopes receive more solar radiation and are consequently drier, making them suitable habitat for drought tolerant species, while cooler north facing slopes are often subjected to a lingering snow pack.

Disturbances such as fire, mudflows, and storms influence both the local atmospheric conditions and the forest. Small-scale disruptions provide unique habitats. A windthrow (a tree downed by wind) creates a sunny spot and often alters the moisture content of the forest floor, creating opportunities for species such as Grand fir, which typically grows slowly beneath the shade of more vigorous associates, while responding quickly to openings created by such phenomena. Rocky, steep slopes may slough off winter snows in periodic avalanches, leaving the surface exposed at an earlier time than the surrounding terrain.

There are many examples of the impact of local climate on trees. Alaska cedar often grows on fog-shrouded ridges, or along wet stream banks with sufficient water. Whitebark pine inhabits rocky areas parched by summer drought, while Pacific silver fir is protected from freezing temperatures within the confines of a heavy snowpack.

Middle elevation snowpack

At low elevation sites, the summer dry spell coupled with local topography is an important factor in tree establishment. Conifers with a quick growth rate, such as Douglas-fir, may withstand the desiccation, but other species with less extensive root systems undergo stress. Pines such as Western white pine and Lodgepole pine often thrive in lowland rocky habitats; by contrast wet streamside environments enable species such as Red alder and Black cottonwood to persist throughout the summer.

In the middle of winter (or spring, for that matter), when weeks may pass without a glimpse of Mount Rainier, and prolonged cold poses more threat than abundant moisture, it is difficult to imagine that trees might ever require water. Yet in sharp contrast to the wet winters, the stress of summer heat and reduced precipitation provides its own challenges to the forest trees.

Altitude

Nestled against the mountain's south flank at over 5,000 feet of elevation, and the destination of choice for thousands of visitors each year, Paradise is a land of striking contrasts. Here, deep winter snows give way to the brilliance of summer flowers, and even the shortest of hikes provides access to flower-colored meadows, with views up long ridges dotted with trees, and down denuded glacial valleys.

Forest Processes

Along the ridges the densely-packed spires define a tree line more than a thousand feet above Paradise. With increasing elevation, the lines thin, and the trees shrink to dwarfed caricatures of the giants below. At the highest reaches, isolated individuals, more shrub than tree, are locked in a struggle for survival at the upper limits of tree establishment.

Trees on ridgeline

The existence of these trees in such an extreme environment may be attributable to several factors. A slight genetic variance that enhances survival may contribute. Or perhaps their establishment occurred at a time when climatic conditions were mild enough to ensure an early snow melt, thus promoting germination in marginal conditions. These seasonal and annual variations are particularly significant in the subalpine zone. If a warming trend persists for several seasons, a group of saplings might survive even with the return of colder times. And, since trees often create a local environment of their own, such success occasionally leads to further germination or cloning. Many Subalpine fir communities in the Paradise area are of a similar age, and there is an observable invasion of meadow areas, suggesting that germination occurred during periods of above average temperatures.

Isolated trees

The isolation of the most remote trees raises questions about the limit of adaptability. Since the glaciers have receded in the current interglacial, undoubtedly the tree line has crept upward, but there must inevitably be a limit to what the trees can tolerate. If so, what determines that limit? And why do the trees grow in such an unprotected location?

Temperature is one limiting factor. For the most part, the variation in the elevation of timberline is latitude dependent. However, winter temperatures are probably not the deciding element for the altitudinal limit, as some species grow in regimes of very low winter temperatures, including extreme arctic and mountainous environments. A minimum growing season (a time when photosynthesis is possible) is mandatory, although this number varies. But an average temperature of 50 degrees Fahrenheit during the warmest month is a good benchmark for minimal growth. This is an approximate rule only; for example, temperature requirements probably differ for initial germination and later growth.

Daily temperature fluctuation during the winter or summer months are important factors. During the cold season, warm air temperatures associated with dry (Chinook) winds kill sensitive exposed foliage; this hazard coupled with moisture inhibiting frozen ground conditions often cause tree death by desiccation. Conversely, summer frosts at a time of growth can not only kill the new shoots but damage the entire tree.

Even in the middle and low altitude forests, where extremes are mitigated, the altitudinal replacement of a species such as Western redcedar, with its relative, Alaska cedar, is undoubtedly influenced by temperature. The sometimes sharp transition from the Pacific silver fir zone to the Subalpine fir and Mountain hemlock forest may owe its origin to the temperature requirements of the silver fir. This demarcation line varies around the mountain's perimeter, perhaps reflecting regional temperature differences.

Lingering winter snows

Temperature effects do not stand in isolation from other components. A micro-climate, created by an established forest, affects the habitat within and around the community. Wind and precipitation vary with altitude, while fire may be caused by unique climatic events, such as high altitude summer storms. Individual survival is always played out in a complex scenario influenced by many non-constant factors.

Aspect

The late spring heat penetrates deep valleys dissected by boisterous glacial rivers. Bursting in response, deciduous trees obliterate the gray of winter. This is a vibrant season, when the air simmers above lowlands colored with the green of new leaves and a profusion of tiny flowers. With the promise of summer in the air, visitors flock to the Park, eager to glimpse the emerging mountain.

The warm day invites a hike, and, attired in shorts and t-shirt, you set out for the ridge, high above the valley floor. Starting within the confines of a dense forest, you know that three thousand feet above the forest will disappear. There, in the higher elevations, a landscape of meadows framed against the backdrop of Mount Rainier, beckons.

"Take an ice ax," advises a friend, but why would anyone need one in this warm forest where the only white color resides in tiny flowers? Ignoring the caution, you begin eagerly, leaning into the effort of climbing, thinking that perhaps there will be time to cross the ridge above, and descend to the other side. There will at least be a time of relaxation in the meadows. Ascending the slope on the twisting trail, the air cools a bit, but not enough to enforce a clothing change. The grade lessens, and with one last effort, the top of the ridge is attained. Turning towards the mountain, you feel the sun beat its radiance upon your back.

The mountain at Van Trump park

Glistening invitingly and appearing close enough to touch, Mount Rainier rises over 8,000 feet above you. Around you are the meadows, colored with flowers, interspersed with dark trees, miniatures of their lowland relatives. The trail is now dotted with a few snow patches, but the benign warmth of the afternoon sun and the beauty of the landscape invite exploration. Shrugging off fatigue, you start an eager descent towards the valley on the other side.

But where is the trail, and where are the meadows? The trees are back, surrounding you with a dense, if reduced in stature, forest, and although the sun reaches the valley below, it no longer warms your back. Snow patches have become a continuous white. As the snow deepens and the trail disappears for good, the ice ax seems not such a bad idea after all, and a direct descent without the aid of a graded trail becomes a reality. Even below, at lesser altitudes, you can see the patches of snow amongst the trees.

What a dramatic contrast to the "other side of the mountain!" Here on the ridge's north side winter lingers, and the icy coating of Mount Rainier now seems a taunting reminder of mountain realities. Shielded from the rays of the southern sun, the trail on this northern slope may not emerge from the snow for a month or more. The snow is deep, and, denied the sun, the plants lay dormant, waiting for a summer made brief for them by location. A chill in the air prods you along, and turning, you ascend to the ridge once more, pause to gaze at the snow-encrusted valley, and descend into the dense forest of the warm southern slope.

Sun

Using a single set of building blocks plants and animals have evolved to occupy nearly every region on Earth. Although both kingdoms depend upon the sun for life, they do so in different ways. Most plants are primary producers, meaning that they derive energy for growth and reproduction directly from the sun. Varying widely in longevity, these rooted beings are adapted to a multitude of environments. Animals are secondary consumers; utilizing plants or other animals for sustenance, they are mobile, conscious, alert, and short-lived. Their dependence upon the plants is ultimately underlain by their common reliance on the sun.

Trees are among the longest-lived plant forms. With little obvious direct interactions, and locked into one location, trees seem oblivious to threat. Nevertheless, as living entities, trees must produce food for their functional parts--the limbs, leaves, branches, bark, and cones. In a similar manner, animals must build and maintain both soft and hard body tissues. For both, these components are composed of organic chemicals which require an external source of energy for production and maintenance. It is in their method of extraction that trees part company with the animals.

Needles — a site for photosynthesis

Trees burn fuel in the photosynthetic process. The source of photosynthetic energy is the sun's visible light spectrum, while the fuel is carbon dioxide, an atmospheric gas. Thus, trees need both the sun and the air. The photosynthetic process is carried out in the cells of leaves and needles where chlorophyll reflects the green wavelength of light from the sun. The quantity of tissue produced is dependent upon the available energy and the efficiency of organic processes within the leaves, making photosynthesis a seasonally variable process. For as the sun moves on a seasonal cycle, the amount of delivered energy changes.

Seasonal fluctuations in available energy are due to the changing angle of the sun. One way to visualize this is to think of the sun's energy as a fixed number of tiny particles enclosed within a tube, like a very long narrow straw. At one end the tube is attached to the sun (the source of the particles), and at the other to the Earth (where the particles are delivered). Each tube at any time holds a constant number of particles, but the surface area of the end of the tube attached to the earth varies depending on the angle of the sun to that particular location. If the sun is directly overhead, as it is at the equator, the circular ends of the tube are of a minimum diameter, and the particles are more densely packed. When the tube is slanted to the surface, such as occurs early or late in the day, or at high latitudes, the end connected to the earth spreads out (as if the straw were cut at an angle), encompassing more area, but containing no more particles than the tube at the equator. Thus, a surface oriented at right angles to the incoming rays receives more energy than one at a more oblique angle.

Tree establishment on slope

Trees employ various strategies to maximize the absorption of this packaged sunlight. Establishment on a hillside of favorable aspect is one approach, and seedlings fortunate enough to germinate in such places often demonstrate a higher survival rate and greater size than those in less desirable locations. Positioning needles or leaves at a optimal angle to the sun's rays is another method for maximizing absorption of energy. This needle configuration may reflect a forest habitat; for example, shade tolerant Western hemlock needles grow at multiple angles along the top of the branches and twigs, effectively covering the twigs and taking advantage of any available overhead light. Douglas-fir surrounds its twigs with uniform nee-

Western hemlock needles

dles, placing this sun-loving species in a position to capture rays from any direction.

By retaining their needles, conifers may photosynthesize year round. Losing their leaves each year deciduous species forgo this method. This strategy reflects a tradeoff between the energy required for the production of leaves each season, and the photosynthetic opportunities available during winter.

Other parameters that affect tree growth are influenced by solar radiation. At Mount Rainier, both the amount and duration of snow are dependent on the sun. A southerly exposure may be an advantageous habitat not only because of the more direct solar energy but because the increased hours of exposure result in a more rapid rate of melting. This factor may be observed in early summer when southern slopes are often free of snow weeks before their northern counterparts. Occasionally, this difference may be great enough to preclude tree establishment on either aspect; the northern because of the lingering snow, and the southern due to excessive dryness!

Lingering snow — Reflection Lake

Different species vary in their requirements for solar radiation. Although all can certainly benefit from direct sunlight, many require it for germination. Others, such as Pacific yew, utilize the filtered light beneath the forest canopy, perhaps reflecting the reduced energy in their slow growth rates. Some species grow slowly but steadily in shaded areas, only to achieve much higher rates in direct sun.

For all species, however, a minimum amount of solar energy is required for photosynthesis and growth. Coupled with geological features, as well as snow and glacial dynamics, this demand is a major factor in the elevation of tree line at Mount Rainier. Seasonally, growth rates are influenced by both summer warmth and winter cold. In addition, species respond differently to available energy. On longer time scales, their ranges are influenced by this variation. For example, during the ice ages, species' distributions expanded or contracted according to their requirements. Today, with global warming, it can be expected that particular species will benefit, expanding their ranges, while others will contract. It would indeed be interesting to return in a thousand years to see the changed forests at Mount Rainier!

Fire

The rain fell in sheets and darkened the old Douglas-fir's red bark, matching it to the black streak along one side. Smelling of charcoal, the blemish rubbed off easily. Remnants of a fiery battle, all of the largest trees were marked with black slashes, while the young trees were free of scars. The evidence and reasons for survival had been obliterated. Only the scars and the small numbers of the large trees were somber reminders that old age had not come easily, and to most it had not come at all.

In the middle of a wet November, fire seems the smallest of forest hazards. But, in late summer, when city sunsets are reddened with the smoke from flames miles away, the role of fire is more apparent. During particularly dry summers, fire threatens constantly, sometimes engulfing thousands of acres. No site is safe. All forests must contend with fire; only the frequency and intensity is variable.

Thus, what may appear to be infrequent from the human perspective is in fact a continuous factor in forest history. Prior to human disturbance, the importance of fire in determining forest composition and health, as well as its influence on evolutionary adaptations, cannot be underestimated. Today, in an environment where humans seek to control fire, the impact on the forest is as certain as in the past, but with markedly different results.

Trees killed by fire

Field investigations of the frequency of fire have shown that there are very few really old forests at Mount Rainier. Over the past 500 years major fires have periodically swept most of the river valleys. Large scale burns occurred on the east side of the Park along the White and Ohanapecosh river drainages. On the south side evidence of burning is found along Stevens Canyon, and in the forests of the Nisqually and Cowlitz Rivers – on the west side, the Puyallup valley burned. Since the arrival of Europeans, at least two major fires have burned. Repeated episodes at particular sites are found, and at least two fires are associated with human activity. Some of the old burns were extensive, the most recent having occurred as little as

350 years ago covered 19 percent of the forested area at the Park!

A few sites have escaped burning for a thousand years or more; most of these are on wet, protected valley bottoms or north facing slopes. Above the Carbon River in the Ipsut Creek drainage, such a forest has survived in a rocky terrain with a northern exposure. The Carbon River valley is home to a 700 year old forest, composed of many species, including some of the largest Douglas-fir at the Park. Similarly, areas in the Ohanapecosh River and White River valleys support old forests. Along the Nisqually, a thousand year old forest grows on an old glacial site at Cougar Rock campground, free from fire since the recession of the glacier.

Bark is a primary defense against fire, and those species most likely to survive a conflagration are covered with the thickest bark. Ancient Douglas-fir trees are covered with a deeply grooved bark five or six inches thick. Old Western hemlock trees also bear a thickened bark. Other species such as Pacific silver fir and Western redcedar have much thinner bark, a strategy that saves resources but leaves the tree more vulnerable. These species are adapted to wet habitats, and thus are arguably less likely to deal with fire frequently. Pacific silver fir forms continuous forests in middle elevations that are snow-covered throughout winter and spring. Probably less prone to lightning strikes than the higher elevations, and more immune to fire in their habitat of deep snows, Pacific silver fir receives some protection from fire in this environment. If fire does occur, it may decimate the fir, and open the landscape to pioneering species.

Pacific silver fir bark

In the summer, unstable air masses rise and set the stage for lightning-induced fire within the patchy subalpine forest. In this region, Engelmann spruce surpasses other associates in height, although it does so slowly. Thus, an Engelmann spruce may be 400 years old. A towering lightning rod, many spruces succumb in a spectacular fashion, split with a shower of sparks ignited by the powerful strike. If it survives, the spruce may bear a long vertical scar.

Fire plays a vital role in forest dynamics. Pioneering shrubs and herbs on sites opened by fire augment the soil with organic material, paving the way for species with different requirements. Fire also clears the accumulation of debris and alters the soil structure. It sets the stage for the regeneration of pioneering species, such as Douglas-fir and Lodgepole pine. In some places, fire may occur so frequently as to ensure a continuing role for pioneering species, but in others be infrequent, thus permitting the development of a mature forest. Fire has undoubtedly influenced such evolutionary adaptations as bark thickness, light tolerance, and rate of growth. Its evidence is everywhere; in the many young forests, in the blackened scars on old survivors, and in the absence of any survivors at all.

Flood

In the cool October that year, rain fell as snow in the mountains, and when the clouds parted, Mount Rainier wore a new white cap. It seemed that winter had come to stay. Clear streams meandered across their wide beds, decorated with leaves from towering cottonwoods and pale alders. A brief autumn was slipping quietly away, or so we believed. But we were wrong, as is so often the case.

The wind grew strong, and a warm rain began to fall. More quickly than what seemed reasonable the new snow melted and flowed into the waiting streams and rivers. Engorged, the rivers swelled and explored their edges with their newly acquired power. Now the rain itself added to the flow, and whipping like the tail of a dragon, the river flung its appendages over the banks and lost its identity.

Flooding along Nisqually River — 2006

Evidence of seasonal flooding is present along glacial streams where dying trees, suffocated by deposits of sand and mud, offer mute testimony to inundation. Scattered about the river bed, and left high and dry after the deluge, uprooted trees indicate the wide range in stream depth and velocity associated with a flooding event. Undercut river banks often retain a layer of topsoil, secured for the present by a tree that hangs over the rocky bed, its dangling roots stripped and exposed by high water. Young Red alders that lined the stream banks poke their trunks through a foot of new sand and mud; only time can determine whether or not they will survive the smothering.

By removing plants and substrate, floods create new habitats, one that favors light-requiring species adapted to sterile, thin soils. Periodic flooding may ensure the continuance of species adapted to these conditions. An example of this is the presence of Red alder along active stream beds. Sites where the disturbance is infrequent may support a community of species that stabilize a previously flood prone area.

Flooding is not necessarily violent in nature but rather may consist of a series of small energy events that create new wetland habitats far removed from the source of flooding. A river may change course during a flooding event, leaving a new environment in the vacated channel or surrounding landscape.

A wide river bed may indicate a flood zone. Not all rivers cut wide channels, but are narrowly confined by geographical features. These flows also contribute to a specialized habitat along their path. Species such as Alaska cedar often grow in the wet and rocky channel edges; Vine maple flourishes in these sites. During a flooding event, transported masses of foliage may take root or sprout new shoots at an entirely different location.

With a climate of yearly rain and snow cycles, flooding plays an important role in forest dynamics. Powerful events, floods provide opportunity in the wake of destruction.

Avalanche

A soundless crack, a slim tear, a moment's hesitation while the mass of rain-soaked snow forged an identity. Submitting blindly to the pull of gravity, the monster gathered itself and began to swell--from above, from beneath, from the sides. Gorging on each bit of loosened snow, ice, and rock, it accelerated forward, accompanied by a rising white cloud of fleeing debris. At breakneck speed, the new entity rose on a cushion of air and flew down the slope and into the valley, leaving in its wake a swath of destruction.

Avalanche chute

Loosened by warmth or heavy snow, an avalanche roars down a slope. On the upper part, bowed trees hold firm, but below, others don't fare as well. Enmeshed in the sliding mass, uprooted trees are swept along, only to come to rest where the grade lessens, where, like pebbles in a stream, they are released from the flow. Unlike deposited stones, however, they lay on top and within, crushed and broken, their branches waving in the air.

Avalanches are a common phenomenon on the steep and snowy terrain of the western Cascades, particularly during the winter and spring. On the steepest slopes, avalanches may be frequent enough to scour the landscape of most vegetation; from a distance these areas appear as giant slashes in the forest, devoid of plant life. At some sites, even the smallest plants are removed by a continual remaking of the landscape, while at other locations, infrequent or ancient slides leave a jumbled landscape of boulders inhospitable to all but the smallest plant life. Many sites reveal a history of infrequent avalanche with a plant community that includes herbs and low-growing deciduous trees; others are of sufficient age and proper conditions to permit coniferous species.

Avalanche chutes

Because of a tendency to bend rather than break beneath the flow, species often relegated to a minor role in other locations may achieve dominance in avalanche areas. At Mount Rainier, Alaska cedar is typically a minor component in middle altitude and subalpine forests. This long-lived species inhabits rocky and wet sites, including avalanche chutes, where its flexible form enables the cedar to withstand sliding snow. Alaska cedar may grow in nearly pure stands, and as the most adapted conifer to these areas, may be an indicator of an avalanche prone slope. Nevertheless, resistant though it is, even Alaska cedar is often broken or uprooted in sites normally free of slides.

Among the deciduous species, Slide alder is sufficiently flexible to withstand some avalanche events. This species is most common in wet areas and steep slopes where it may share the terrain with Vine maple, another species typically relegated to a small understory role in the forest.

Avalanche paths provide a continual challenge. Tree may reach reproductive status, only to be swept away, diminishing the opportunity for further colonization. Some may survive if sufficiently flexible to withstand breakage, or capable of enduring beneath the crushing load. A slow growth rate may be advantageous in a situation where substrate sterility constrains growth. In the end, survival may depend on slight differences in form, on location, or on the presence of other trees.

Not all avalanche paths are apparent. Bright gashes mark steep slopes regularly swept clean, while other sites subject to infrequent avalanches are more likely to support an encroaching forest. In these areas the trees may not only alter the site to the extent that it is not recognizable as an avalanche path, but affect the route and magnitude of an avalanche should it occur. The effect of an infrequent avalanche or one of small magnitude may be to preclude total decimation; the quick recovery of this forest will continue to affect avalanche dynamics.

A common disturbance force at Mount Rainier, avalanches are prohibitive events for the majority of tree species. Pioneering opportunities are provided for a few resistant conifers, and communities of deciduous shrubs and trees often thrive where others are excluded. Old age is rare in these landscapes, but in time, even avalanche chutes may be altered by resident species.

Wind

Protected from the cold air above by snow pack, the buried seedling lay dormant. Nearby, a taller neighbor battled the winter storms half in and half out, its lower trunk and roots in the sheltered zone, its foliated top exposed to the cold air.

In the Northwest, winter weather is often a mixture of snow and cold rain alternating with warm southerly winds that rise quickly, sweeping through lowlands and mountains. These varying conditions create freezing and melting cycles, stressing the trees. The warm periods may spur premature growth, leaving the tree susceptible to damage in the freezing cycle. Strong winds often desiccate and kill coniferous trees. Without the storms, however, the trees might not survive at all, for it is the accompanying snows at higher elevations which provide shelter through the cold winter and ensure continuing moisture during the dry summer.

Trees pruned by wind and snow

An efficient pruning force, wind removes weakened trees, creating openings in the forest. Established trees and other plants formerly relegated to understory roles may undergo accelerated growth in such locations. The fallen trees are not wasted but serve as sources of organic material to the soil, and as a host to new seedlings.

Trees are not defenseless against the wind's force. Deep root systems are an effective strategy, and yet this characteristic seems the exception rather than the rule. And extensive roots are probably more prevalent as compensation for drought than wind. Thus, lacking a deep root system, many species are prone to windthrow, including Pacific silver fir and Western hemlock. Engelmann spruce, an inhabitant of diverse, cold environments, is also shallow-rooted. Perhaps under conditions of consistent moisture, there is little need for a deep, stabilizing taproot.

On occasion, hurricane-force winds may decimate forests. However, even in this event, many trees may be protected in densely packed communities that ameliorates the wind's force. Along the edges of the forest, trees are more likely to be downed.

In wind-swept landscapes, trees are often contorted in shape. Since trees cannot move, survival may depend on this versatility to take on more than one form. With their growth rate slowed, these trees hold fast to the rocky substrate.

Alaska cedar on ridge

Wind is strongly coupled with other factors. Storms may bring sufficient wind and rain to cause snow and mud avalanches. Summer heat is ameliorated by mountain breezes. Transporting moisture-laden fogs on updrafts along mountain slopes, the breeze may promote ridge top species, such as Alaska cedar. Wind creates openings for pioneering species. In the spring, evidence of the wind's passing may be witnessed in an altered landscape, a pile of broken trunks and branches, or a new course cut by a stream. On a windy day, a cloud cap on the mountain provides a reminder of this ever present force.

Eruption

Like bubbles in a simmering pot, the Cascade volcanoes protrude from a caldron of submerged lava. St. Helens, Adams, Rainier--their names reflect European history, but the Native Americans were the first people acquainted with their fiery nature. Greatest of all was Tahoma—Rainier--a fractured and worn mountain. But fire still smoldered within, and the recalcitrant Tahoma simmered and occasionally burst, tossing ash and cinders, and sending cascades of mud and ice down remade slopes.

A phenomenon of the current ice age, the Pleistocene, Mount Rainier is young as mountains go. Composed of multiple layers, the mountain is a strato-volcano with an explosive history, into which the eruption of Mount St. Helens in 1980 provides some insight. Erupting from a flat, granitic plain, the birth throes were spasmodic and immense, as layer upon layer of molten rock and debris was added to the mountain. Pelting the landscape with rocky debris, and spewing sticky lava down eroding flanks, over thousands of years a volcanic cone was pushed skyward to an elevation that was, at one time, at least a thousand feet higher than today.

Little Tahoma

As the eruptions continued to fill and remake the terrain, giant ice age glaciers carved away at the mountain's flanks, while water and wind contributed to the structuring of a varied landscape. Composed of a mixture of deposition, glacial outwash, and mud, rock, and ice flows, the terrain is characterized by steep valleys and escarpments contrasted with plateaus, glacial deposits, and wide lowland river valleys. Ridges of various ages extend spoke-like from Mount Rainier, while large, park-like lands owe their origins to giant eruptions and subsequent erosion.

Life and death is played out within the context of this eruptive and climatic history. Giant flows down the mountain's flanks have periodically inundated forests, enabling the establishment of pioneering species adapted to the volcanic soils. Glacial advances also swallowed whole landscapes. Today, remnants of those glaciers cloak the mountain and extend into the valleys. Here they influence the local weather and provide water for plants and animals alike. The parent material of the forest soils is a product of the climatic and geologic effects. Following periods of active volcanism species adapted to thin, well-drained soils were at an advantage, while renewed volcanic or glacial activity ensured a continuing presence for these pioneers. On the other hand, the breakdown and alteration of the raw material during quiet periods pro-

Tree establishment on glacial moraine

moted other species.

This environment of change and stability explains in part the number of species found at the mountain. For example, Whitebark pine and Lodgepole pine pioneer lava fields with a freshly erupted appearance. At high elevations, lodgepole sometimes grows in nearly pure stands on rocky, well-drained terrain. This cosmopolitan species is also adapted to the hot spring environment snear the mountain, while another pine, Western white pine, grows on rocky outcroppings. Lowland species such as Douglas-fir and Western hemlock alter the volcanic soils also, achieving great size in the valleys and surrounding hillsides, while deciduous species such as Black cottonwood and Red alder pioneer the banks of glacial rivers. Vast forests of Pacific silver fir grow on the volcanic rubble of the middle elevations, and Noble fir forests populate steep south-facing hillsides swept clean by glaciers.

Pioneering species along Carbon River

Young forests growing on fresh lava or de-glaciated land are typically sparse in appearance. Where the breakdown of rock and persists, forests, often composed of new species, thrive. Old lowland forests are the most complex and productive of all. In addition, the range in elevation resulting from the new mountain excluded many species from their former range, while permitting the establishment of those species adapted to a colder regime. With a cessation in eruption, invasion of a multitude of species became possible.

At the summit of Mount Rainier, remnants of volcanism are evident in warm caves. Approximately 5,000 years ago, an increase in activity triggered an enormous mud flow that changed the landscape and impacted human communities nearby. Fifty years ago climatic pulses produced a flow that obliterated a forest whose remains are still visible. Blinks of an eye in geologic time, these events serve as a reminder that Mount Rainier is an active mountain.

Human Impact

The signs of human impact are not as apparent at Mount Rainier as they are in much of the Pacific Northwest. Outside of the Park, the forest has been altered to such an extent that the sight of denuded hills and mountains is as common as old forests were a hundred years ago. Most people do not realize the magnitude of the clearing; many do not comprehend that Douglas-firs comparable to the size of the giants that line the Carbon River road were by no means unusual prior to the advent of extensive logging. The Northwest is a logged landscape; the lowland old growth forest only a memory, and the forests of Mount Rainier remnants of what once was.

Mount Rainier National Park includes over 240,000 acres of which approximately 60 percent is forested; much of this is middle and high elevation forest. The landscape has, in the past, been swept by fires; the current policy is to suppress fire, a practice with an undetermined impact. Not all fires have been due to natural causes; sheep herders set fire to maintain meadows, and recent human activity has been the cause of at least one forest fire, that near Longmire in 1894. Possibly a slash fire that burned high elevation habitat near the north Puyallup River in November of 1930 owed its origins to human sources. Many whitened snags mark the site, and reforestation has been a slow process.

Longmire Hot Springs

Selective logging has left its traces at many locations where buildings or roads were constructed. At Longmire, clearing of the land around the hot springs area and the construction of several buildings, including a hotel, to accommodate the Longmire family and tourists, was extensive during the 1890s. In particular, the area enjoyed great usage following the establishment of the Park in 1899. Other buildings at the Park are used extensively, including the lodge and other buildings at Paradise and Sunrise. Campgrounds were more widely utilized in the past; today White River and Cougar Rock accommodate campers in varied landscapes.

Mount Rainier is a hiker's paradise, and miles of trails, including the Wonderland Trail which circumnavigates the mountain, provide opportunity for access. At the major visitor areas, Paradise and Sunrise, extensive trail use and incursion into the fragile meadows has necessitated the construction of wide and sturdy trails and visitor restrictions. Trails that access the backcountry enjoy less use, but these often enter fragile meadowlands that are particularly prone to disturbance.

Roads at Mount Rainier are good locations to observe the effects of disturbance on local flora; the major highways in particular cut broad paths which provide opportunities for species that might otherwise be excluded. For example, Douglas-fir saplings thrive in the open areas along the road to Sunrise. Herbs otherwise more restricted in range may grow in profusion along Park roads; step a few feet from the road and an entirely different plant community is typical. Old Park roads provide their own unique set of conditions; an interesting example of this is a site paralleling the Nisqually River above Kautz Creek. Many saplings find an opportunity here, including Western hemlock, Western redcedar, and even a Western white pine, while herbaceous plants thrive in a much thicker profusion than found only a few feet away.

Kautz Creek

Where the landscape is dominated by the immensity of the aging volcano, and the forests quickly swallow the trail-bound hiker, human impact may seem minimal. Perhaps it is more important to consider what the forests at the Park have become with respect to the surrounding landscape. Mount Rainier is an island, not only because its bulk rises above a ring of forest, but because the forests themselves are an island. On every side, along every mile of the perimeter, human activity makes itself felt. Outside the Park, old forests are fading memories, and the scope of this alteration and its impact on isolated regions like Mount Rainier is difficult to assess. That there is an impact is undeniable; that the magnitude of the change may make itself felt in years to come, is certain.

Glaciers

The planet is in the ice ages, those pulsating geophysical events that have molded the flora, fauna, and geology for millions of years.

We live in the Pleistocene, an ice age defined by periodic advances of immense ice sheets over much of the Northern Hemisphere, and rapidly shrinking remnants during retreats (the interglacials). Thousands of years old but young by geological standards, the current interglacial is a time of diminution, an era when the spectacular glaciers that define the topography of Mount Rainier are slim reminders of the ancient sheets. These small glaciers provide clues to the magnitude of change inflicted by their diminished predecessors and continue to exert a strong influence over the biological and physical environment.

The relentless icy rivers have prematurely aged Mount Rainier, altering the youthful appearance of a blackened volcanic cone to a gnarled and whitened caricature of youth. Today, twenty-three glaciers carve Mount Rainier, and, radiating spoke-like from their snouts, glacial rivers extend the influence to the lowlands down long and wide valleys.

The receding Tahoma Glacier

The terminus of a particular glacier fluctuates in response to climate. For the most part, the glaciers at Mount Rainier have receded during this century, and, in so doing, have deposited markers of their passing in lateral and terminal moraines, mounds of debris that are indicators of glacial age. Tree establishment is not far behind the recession of a glacier, and seedlings may sprout on glacial debris in as little time as a year after final melting. Facilitating the aging of the glacier, the saplings grow amidst the rocks and ruin of what appears as the most hostile terrain of all, a pile of rock and sand stripped of soil and fertility. It is, however, the barrenness that permits easy invasion.

The alders are among the first trees to pioneer deglaciated land. Their ability to fix atmospheric nitrogen makes them especially beneficial to the soil building process, and their presence helps to stabilize stream banks and steep hillsides. Both Red alder and Slide alder grow in glacial areas, with Red alder most common at lower elevations, and Slide alder in higher regions. Conifers adapted to rocky terrain, such as Lodgepole pine and Englemann spruce (on the north side of the mountain) soon invade. In time, other coniferous species, such as Western hemlock in the lower re-

Pioneering species along Nisqually River

gions, and Pacific silver fir in the higher, continue the reforestation process, nearly obscuring the glacial moraines.

Much of the soil at the Park originates from colluvial material, an unconsolidated mixture of glacial debris and volcanic deposits. Alluvial depositions are common in major river valleys; this mixture of sands and gravels owes its origin to glacial floods and outwash. Thus, soils formed from these parent materials are geologically recent constructions. Their composition is determined by the breaking down and re-working of these materials by both biological and physical processes. The soil structure influences the plant and animal communities, while they alter the soil itself.

Glacial till — Winthrop Glacier

Although the current interglacial is a time of minimal ice expanse, at some future time a swing towards increased glaciation may occur. The forests will respond to climate alterations that are both the cause and effect of the fluctuating ice sheets, and the composition will of necessity be altered to one compatible with those changing conditions. Continued opportunity will be provided to those disturbance species capable of inhabitation of glacial till, while the cold will promote expansion of some species' ranges and limit others. What appears as stability within the context of our brief life span is in fact, only a snap shot of a system that is continuously subject to modification. Directly and indirectly influenced by glacial forces, waves of change will continue to flow through the forest.

Biological and Physical Processes – In Combination

Ramrod straight, the densely packed young trees capture the sun's energy in their needled crowns. At the ground, a sparse understory attests to reduced sunlight. By contrast, a mixed forest that ranges from saplings to giants six feet in diameter supports a profusion of shrubs and herbs beneath the irregular canopy. The two forests appear as different as a plantation is from a jumbled garden. But it is difficult to determine precisely what those differences are. Does the old forest support more plant and animal species? Are the interactions more complex, the competition for living space more or less keen?

As a young forest ages, physical factors such as wind and available sunlight bring changes. Weakened trees fall, adding humus to the forest floor, while new species invade the opened terrain. Fire may sweep the forest, removing the understory while leaving survivors behind. Pioneering species take advantage of the altered landscape, but as time passes, more shade-tolerant species are favored. The forest is now maturing to a "climax stage," a term that implies long term stability .

Fire and regeneration

However, an aged forest is not a fixed entity, but rather exhibits a temporal and spatial mosaic. Differences in longevity alone imply alterations while factors such as climate change further affect the composition of the forest. In these cases, the definition of ancient becomes problematic, for in the absence of obvious disturbances, such as fire, a forest may continue to appear stable although it undergoes a change in member species.

Ancient forests may be lost in an instant, such as in the case of a mud flow. An event of this magnitude opens a whole new scenario, replacing the old forest with a young community. On particularly impoverished terrain, such as glacially altered sites, tree establishment may be delayed for years, and will proceed in a habitat defined by herbaceous predecessors. Eventually even these sites yield to aging forests, as the trees create shade for seedlings particularly sensitive to drought, enhancing the environment for slower growing but shade tolerant species. Some sites may change so dramatically as to affect forest composition on a very long term. Turf building species may contribute material sufficiently thick to fill a wetland, creating a new habitat for species intolerant of inundation. In other regions, where a river changes course, the process is reversed, and new wetlands permit the establishment

Wet meadow with Skunk cabbage

of alders and cedars otherwise excluded from a mature forest environment. These erosive processes often deposit clumps of material within a stream bed; here species may gain a foothold, stabilizing the new substrate and influencing the river's course.

Within an old forest, death and decay provide a continuous source of material, supporting a multitude of fungi and bacteria species. These alter the soil, rendering it more hospitable for some species while excluding others. In addition, many species of plants and animals make homes in the branches and crowns of aging trees.

The effect of elevation on solar energy is a particularly important factor in forest composition. Not only does altitude affect the potential size for a particular species, but it undoubtedly has influenced the speciation within a family as well. An example of this is found in the true firs, the *Abies* genus.

Grand fir — White River

Four true fir species cover the entire forested altitudinal range at Mount Rainier. Limited at the Park to altitudes between 2,000 and 3,000 feet, Grand fir grows most of the major river valleys (although it is absent in the Ohanapecosh drainage) along exposed sandy banks. Grand fir and Pacific silver fir are associates in a narrow elevational band, while in the subalpine regime, Pacific silver fir mingles with Subalpine fir, the most common species of this habitat.

Grand fir may grow to a diameter of four feet or more; Pacific silver fir typically reaches at most 30 inches, and Subalpine fir rarely exceeds two feet in diameter. Subalpine fir and Pacific silver fir inhabit the subalpine zone at altitudes ranging from 5,000 to greater than 6,000 feet, but the silver fir is typically greater in diameter and height. Where species grow together size disparity is typically still maintained.

Potentially the largest of the true firs, Noble fir is a bit anomalous, however. Larger than Grand fir, Noble fir may reach truly large proportions on disturbance sites between 2,000 to over 5,000 feet. Noble fir often reaches a diameter of more than four feet— sometimes much greater —and a height greater than 200 feet, with an average volume 100 times that of Subalpine fir. At low elevations, Subalpine fir may reach 100 feet in height, but at Mount Rainier, where it is typically confined to high elevations, this species grows a quarter that tall and barely a foot in diameter.

The four true fir species inhabit forests from the lowland to subalpine regimes. Other groups, too, encompass this altitudinal range--these include the spruces (Sitka and Engelmann), cedars, and pines. As in the case of the firs, the lowland Sitka spruce sustains a much greater growth rate than Engelmann. Western white pine and Lodgepole pine often grow in close proximity, but the white pine grows to much larger size.

Noble fir

Douglas-fir and Western hemlock attain their greatest size as lowland species, too, although both grow at altitudes in excess of 4,500 feet. Although reduced in stature at the higher elevations, Douglas-fir typically grows to greater size than its associates. Thus, these large species, adapted to the lowest altitudes, would seem to have a great competitive advantage in the higher elevations, but here they are clearly outnumbered. Western hemlock no longer is the climax species, but is replaced by Pacific silver fir, and Douglas-fir maintains a marginal presence.

Thus, associates in a particular habitat sustain growth rates in response to the interplay between genetic and external parameters. This interaction may also be observed in the wide range of form and stature acquired by some species, while others show less obvious response to the environment. And although the size and age of an individual tree reveals something about the habitat, small differences in local conditions can make this factor difficult to determine. Particularly in the subalpine environment, trees separated by small distances may vary considerably. In many locations, soil fertility and moisture contribute to similar variations, and local openings influence growth rate.

An individual forest depends on a whole suite of factors, and no single species within any group fills the entire range of available environments. Differences in moisture, temperature, and substrate requirements reflect past environments, the limits of adaptability, and the dynamics of species interaction. Acting within evolutionary constraints and ecological history, the resident tree species at Mount Rainier demonstrate an adaptability that is reflected in their growth rate, potential age, and size.

The Forest Revisited

Dauntingly close, the mountain appeared to loom overhead. Only the specks of climbers on the fractured glacier served as reminders of the immensity of the icy face. Far below the summit, at 7400 feet of elevation, breathing was laborious, and the sun barely warmed the chill of the mountain wind.

A sparse herbaceous layer struggled to make soil of the porous volcanic debris; there was no forest here, no rich understory, no tall trees. And yet near this location an Engelmann spruce supposedly grew near the uppermost limits of the tree line, if only we could find it. Nothing appeared to acquire more than a few inches of height, and the last stretch of recognizable forest lay a thousand feet below.

Engelmann spruce — Third Burroughs

Was it possible for any tree to grow at this elevation, so close to the mountain that even in the middle of summer one could feel its icy breath? Close enough from the perilous ridge carved by a glacier that stretched below, its crevasses a reminder of the power of moving ice? And yet here it was, a much-diminished spruce, blasted by the elements, an unwitting inhabitant of the windswept terrain.

Tree Islands

At the highest elevations on Mount Rainier's north and east sides Engelmann spruce, Whitebark pine, and Lodgepole pine are the major tree players in the forest community, if the sparse numbers could be said to constitute a forest at all. Certainly they do not define a community in the same sense as in the lowlands. Often present as solitary trees, sometimes as small stands, these species create their own micro-habitat. More continuous forest is found at lower elevations, particularly on south facing slopes. The most common tree in the subalpine region is Subalpine fir; this is the species that may also grow in tree "islands" of multiply-aged trees. Mountain hemlock often attains considerable size, but it is much less common throughout the Park. Few deciduous species can survive the higher altitudes, but willows and Slide alder grow in wet and steep areas. It is not unusual to find severely stunted trees here, their age difficult to determine.

On the south side of the mountain, Subalpine fir dominates the higher elevations. The most common species of the middle altitude forest, Pacific silver fir, may also grow near timberline. Relegated to a minor forest role, Mountain hemlock is reduced in size in these forests. Absent on this side are Whitebark pine and Engelmann spruce; these are restricted to high elevations on the mountain's north and east side.

In terms of species, the greatest number are found in the middle elevations. From the lowland forest conifers, Western hemlock, Douglas-fir, and to a lesser extent, Western redcedar, are present at this upper end of their range. Most of the deciduous species are confined to the lowlands, but Vine maple grows in avalanche zones. The whole set of middle altitude conifers are found here, including Alaska cedar, Mountain hemlock, Noble fir, Western white pine, as well as the dominant Pacific silver fir.

The middle altitude Pacific silver fir forests may take on a bluish cast from the combination of deep green needles and white stomatal lines on the underside of the silver fir's needles. The trunks are straight and pale gray in the younger trees and darker gray with narrow groves in the older. Here also Alaska cedar is a distinctive species with its droopy form, greenish-yellow foliage, and strips of gray bark. Found on rocky prominences above streams, this species occasionally forms a continuous, dense mass of shrubbery in avalanche-prone terrain. The larger Western redcedar grows to above 4000 feet elevation, but is uncommon by comparison to the other species, and is typically found in wetter sites.

Alaska cedar

Young Noble fir trees have soft, thin and bluish needles with a characteristic "hook," while the mature trees bear needles that are thick and brushed up, giving the tree a very stiff appearance. Nobles are unmistakable. They are ramrod straight, covered in beautiful maroon-tinged platy bark, bear large (six inch) erect cones near the tops of the trees while the crown is often confined to the top third or less of the trunk. This is a disturbance species with demanding light requirements, but forests dominated by Noble fir do occur in the middle elevations. Although never as common as Pacific silver fir or Western hemlock, Noble fir may share the disturbance role with Douglas-fir .

Today, the scarcity of old lowland forests outside of the Park makes those at Mount Rainier a rare treasure. Although smaller by comparison to many long since felled by the ax, the oldest lowland forests at Mount Rainier are populated with large trees. Six and seven foot diameter Douglas-firs, four foot Western hemlocks, old Western redcedar groves--these are some of the inhabitants of these uncommon forests. Beneath the giants, Vine maple and Pacific Yew make occasional appearances; along the streams and rivers, great communities of Red alder transform the infertile rock and sand to soil fit for the other species. The alder shares the habitat with the largest deciduous trees, Black cottonwood, and the shorter if equally massive Bigleaf maple. Sandy, open banks along glacial rivers provide habitat for grand fir, an enigmatic species found more commonly east of the Cascades. And diminutive willows thrive unobtrusively in the wettest habitats, both in the low elevation forests and above.

Old lowland forests are usually dominated by Western hemlock while Pacific silver becomes more common above an elevation of 3,500 feet or so; this fir in turn yields to Subalpine fir at the highest altitudes. Where fire has altered but not totally decimated a forest, a multiply-aged stand of more than one species creates a diverse community characterized by openings and a rich and varied understory.

Subalpine fir forest

Many avalanche-prone areas support a tree community, including Slide alder, Vine maple, the willows, and Alaska cedar. These species also play an important role in de-glaciated areas, paving the way for a forest composed of different species. Old glacial sites now support young or middle aged forest communities. Areas where avalanche or mud flows occur provide opportunity for many opportunistic species, extending the range of some while ensuring, for the time being, the continuing presence of others.

There are many words used to describe the state of the forest. Disturbed, climax, ancient, old growth--all of these terms apply in describing forests of varying age. A "climax" forest is generally defined as one where the regenerating species are the same as the older members of the community. An example of this forest type at Mount Rainier is the middle elevation Pacific silver fir forest. Three foot diameter old trees form a canopy over younger members of varying ages. But climax forests are not unchanging ones. Even those forests which rarely undergo disturbance are subject to variation. The death of an old tree may provide opportunity for species previously excluded, and old forests may undergo a slow change in composition not immediately obvious to the observer. Thus, across the landscape a flux or wave moves of varying times and duration.

The oldest lower elevation forests at Mount Rainier may date back a thousand years, but they are not necessarily complete in terms of species stability. Often these forests, such as the beautiful forests along the Carbon River road, are in a slow transition, as the old Douglas-firs slowly yield to younger hemlocks and spruce. Influenced by the life and death of the old and young trees the understory of this forest is also subject to change.

Lowland forest

In many ways, the forests of Mount Rainier are similar to other forests throughout the Cascades and Olympics. Nevertheless, the position of the mountain on the western boundary of the Cascades, its glacial history, its composition and geological youth greatly influence the forests and have provided some unique characteristics. Fire has been a regular disturbance factor, one whose influence is witnessed everywhere, and it may be surprising to learn how few truly old forests are found at the Park. Perhaps the most amazing attribute of all is the relative freedom from human disturbance, although not human influence. Surrounded by forests altered by new demands and expectations, the forests of Mount Rainier provide a glimpse into a more distance past, and the chances and the hopes of an uncertain future.

Middle elevation forest

At Mount Rainier, Pacific silver fir is the most commonly encountered tree species between elevations of 3,500 and 5,500 feet. Sometimes also growing near timberline above 6,000 feet, in the lower elevations Pacific silver fir saplings typically dominate the shady understory.

The shade tolerance of Pacific silver fir may be due in part to the large surface area in the densely packed horizontal and upturned needles. Configured in a two-rank horizontal row that effectively conceals the twigs, the tops of the needles are glossy-green. Turn the needle over, and two white lines cast a pale silver glow to the slightly upturned branches, giving the tree a bluish look.

Pacific silver fir needles

The upward-curved needles on the higher branches resemble those of both Noble fir and Subalpine fir. During the spring and summer, three-to-five-inch long pale green female cones stand erect near the top of the tree. Turning purplish in the fall, and covered with pitch, the cones dry and release the heavy seeds. Disintegrating in place, only dark spikes are left as reminders of their presence. Like many other conifers, dissemination of the seeds depends upon the variable winds.

Pacific silver fir trees may grow to a height of 150 feet or more, with the trunk typically no more three-and-a-half feet in diameter. There are large individuals, however, at the mountain. In the Carbon River valley the author has measured a Pacific silver fir at over eighteen feet in circumference (six foot diameter).

This species thrives in regions of heavy snows, and is often the only reproducing species in this habitat. Here the seeds must germinate quickly, often before the snow completely melts. Mortality rate due to late freezes is undoubtedly high.

In subalpine forests, Pacific silver fir plays a minor role. This may be due to the cold, or deep snows, or perhaps an inability to reproduce vegetatively.

Shallow-rooted and thin-barked, wind or fire may spell instant death for young and mature trees alike. In the absence of these large-scale forces, individuals eventually succumb to attack by fungi or insects. With a potential life span of 500 years, and an adaptability to snow and cold, Pacific silver fir has attained a long term niche in Northwest forests.

Reflections

Death

The buzz of the saw ceased, and a motionless silence fused the air with tension. A flicker of movement shimmered at the corner of our vision, but still the quiet. In anticipation we waited, and then, looking upward, sought the source of discordant sounds now beginning to coalesce into a crescendo. Synchronized with the intensity of the scene above, the sound revealed its source. Ripping and tearing, a giant gnashed the forest with branched teeth, and, with a thrust of its immense trunk across the floor, and, ending its fall before we knew it had begun, completed its own destruction with a shudder. It lay near our feet now, branches askew, flipped 90 degrees, now prostrated in a mute offering that failed to show humility. Stretched into the darkening forest, the felled monolith displayed itself, its pieces and parts still quivering, only moments before out of our reach. We were invaders at a private funeral, a ceremony for a monster which had breathed far above us, a giver of life to sky creatures, now suffocating on the ground. Disturbed by the destruction in a place where we did not belong, we fled this fractured landscape and sought solitude in a more ageless land.

Cadence

The wind catches the treetop branches, and, flowing through and around them, imparts a swaying motion to the lacy appendages. The flexible trunk bends a little, yielding just enough to resist breaking while preserving an appearance of stolid indifference to the forces around it. Closer to the ground, the trunk remains immobile, unaffected and sheltered within the forest community. No longer free to move at will, the headstrong wind is transformed to a gentle breeze, reigned in by the mass of trunks and foliage. The trees have created an environment suitable for a multitude of life forms.

Just for today, the observer abandons the quest to understand, to dissect the beauty into a multitude of parts, to apply the rigors of the disciplined mind. She drops to her knees on the forest floor, and runs her hand through the soft mosses. Gazing upward, she sees the swaying treetops, and nestles her back against an untouched trunk. Knowing that the forest is indifferent to her presence floods her senses with comfort rather than alarm. There is room for her here, an unstated invitation from an encompassing host. Her eyes close, and her notebook drops to the ground.

Color

The dark winter trees bend somberly beneath their snowy load, brightened on rare sunny days by their white-covered tips. Stark limbs of summer's forgotten green trees frame the ashen skies, and even the rivers slow in this quiet land, diminished by the cold. The whole forest seems to sigh, forced to withstand the black storms of winter in anticipation of warmer days.

The pale red glow of alder catkins signal spring's beginning, and white Bitter cherry blossoms brighten the scene. Life turns riotous as buds burst under the pressure of their eager green contents, minute cones lengthen in anticipation, and water rushes everywhere. Soon the snow will be gone, and the wet forest floor will turn pale and dusty. The brief summer has come with subtle color replacing the gray. Trees sway gently against the bright sky, and push their limits of growth while staking their future on the black seeds buried in darkening cones and summer flowers.

Bright green tree tips darken and harden with the advancing summer days, and autumn comes, sometimes quickly, sometimes with deliberate subtleness. The lowland forests and avalanche chutes are shot full with red and golden hues, as the Vine maple relinquish their understory position for a center stage show. Purple cones dot the treetops of the Subalpine firs, and Black cottonwoods defy their name with yellow crowns. Muted brown leaves line the river bank, while tiny cones spill their contents on the forest floor, and the white cotton that filled the springtime air lays quiet, relinquishing its hopeful seeds. The seasons begin their cycle once again.

Immensity

I tightened my boots, adjusted the old pack, and glanced skyward, thinking wistfully that a patch of blue would do much to relieve a monotonously ashen sky. The frequent looks didn't seem to help; if anything, the sky swung lower and darkened more, or perhaps I was approaching it with each footstep. Ignoring the gray and lowering my eyes to a more varied scene, I gathered into view the ancient Douglas-firs which had drawn me to the Green Lake trail that day. Mute, dark forms, they stretched skyward, and, for the moment, even they seemed to add to the oppressive atmosphere. What a different feeling than a hike on a sunny ridge!

But I was committed to the hike now, and as my distance from the river valley increased, I felt a growing interest in the silent forest. It was absorbing, and my senses heightened even as my body relaxed. I swung along, expecting an encounter, but not seeking it. The magnitude of the meeting came as a surprise.

The trail narrowed as I rounded a bend, and, absorbed in the small details of this dark forest, I nearly missed it. A massive red form, nine feet wide and taller than any wall I had ever seen, rose beside the trail. I gazed upward, seeking the top, searching for the top. Filtered by the sun, filmy branches appeared black against a sky that seemed a million miles away. I caught my breath.

A giant Douglas-fir, alone, as if its size had bestowed a deserved isolation. Only the footprints of curious humans surrounded it; even the plants had withdrawn. It stood touched but undamaged, memories of youthful competition obliterated by old age. If it needed to prove its worth, that task was long since accomplished--it was silent, immune to the external world, answering only to an internal clock ticking the slow onset of death, which, until finalized, could not be observed. The tree was alive and well.

It was colossal and immeasurably tall. With an entire life passed on this tabled land, the fir had lived through perhaps a thousand summers, dispersed countless seeds from numberless cones, felt its thick bark licked and blackened by fire, and stood firm while others fell, scattering their remains around its silent form.

Was it by chance only that this old fir stood, alone, a relic of a lost time? Was it favored by the elements, or was some other force responsible, a special characteristic of the young sapling centuries ago? What combination of factors had bestowed this quiet old age?

The lonely giant didn't need me. I backed away, gazed upwards once more at the dark branches softened by the misty sky, and turned away. Subjected to the old age of living things as surely as my humanity subjected me, it was I who needed the connection. I would come again another day, quietly – expectantly.

Appendix A
The Species

Conifers

Pacific silver fir — *Abies amabilis*

At Mount Rainier, Pacific silver fir is the most commonly encountered tree species between elevations of 3,500 and 5,500 feet. Sometimes also growing near timberline above 6,000 feet, in the lower elevations Pacific silver fir saplings typically dominate the shady understory.

The shade tolerance of Pacific silver fir may be due in part to the large surface area in the densely packed horizontal and upturned needles. Configured in a two-rank horizontal row that effectively conceals the twigs, the tops of the needles are glossy-green. Turn the needle over, and two white lines cast a pale silver glow to the slightly upturned branches, giving the tree a bluish look.

Needles along tops of twigs

The upward-curved needles on the higher branches resemble those of both Noble fir and Subalpine fir. During the spring and summer, three-to-five-inch long pale green female cones stand erect near the top of the tree. Turning purplish in the fall, and covered with pitch, the cones dry and release the heavy seeds. Disintegrating in place, only dark spikes are left as reminders of their presence. Like many other conifers, dissemination of the seeds depends upon the variable winds.

Female cone

Pacific silver fir trees may grow to a height of 150 feet or more, with the trunk typically no more three-and-a-half feet in diameter. There are large individuals, however, at the mountain. In the Carbon River valley the author has measured a Pacific silver fir at over eighteen feet in circumference (six foot diameter).

Large seeds

This species thrives in regions of heavy snows, and is often the only reproducing species in this habitat. Here the seeds must germinate quickly, often before the snow completely melts. Mortality rate due to late freezes is undoubtedly high.

Male cones

In subalpine forests, Pacific silver fir plays a minor role. This may be due to the cold, or deep snows, or perhaps an inability to reproduce vegetatively.

Shallow-rooted and thin-barked, wind or fire may spell instant death for young and mature trees alike. In the absence of these large-scale forces, individuals eventually succumb to attack by fungi or insects. With a potential life span of 500 years, and an adaptability to snow and cold, Pacific silver fir has attained a long term niche in Northwest forests.

Grand fir -- *Abies grandis*

A coastal species typically confined to elevations below 2,000 feet in the western Cascades, east of the crest Grand fir inhabits wet regimes between 3,000 and 6,000 feet. This versatile species also thrives on lowland banks along the glacial rivers that trace giant spoke-like paths through wide valleys. Although the proximity of the rivers poses a constant threat, healthy saplings populate the sandy and rocky banks where they compete successfully with Douglas-fir, an associate of open environments. Old Grand firs, their trunks turned pale with lime-colored lichen, attest to a historical presence of the species at the Park.

Glossy needles

Grand fir's long, glossy-green needles spread at nearly right angles from narrow twigs; turning the needles over reveals two white bands beneath.

The sweeping branches of young trees appear skirt-like, while old individuals develop a domelike top, with the lowest branches well out of reach. The grooved, gray bark of mature trees resembles that of Pacific silver fir, a relative of middle and subalpine forests. During the autumn, two-to-four inch cones stand erectly near the tree tops, confirming the true fir lineage.

Grand fir grows on soils ranging from shallow volcanic-based tuff

Female cones

to deep alluvial substrates to shallow, sandy soils. It is an opportunistic species, filling climax or intermediate roles, although it is typically found in mixed stands. Moderately shade tolerant, Grand fir may exist in a suppressed state beneath the forest canopy, but will grow quickly in full sun, sometimes achieving a growth rate equal to that of Douglas-fir.

Depending on the availability of soil moisture, Grand fir saplings may establish extensive or shallow root systems. This flexibility impacts the drought tolerance of the seedlings, permitting survival in as little as 25 inches of rain a year.

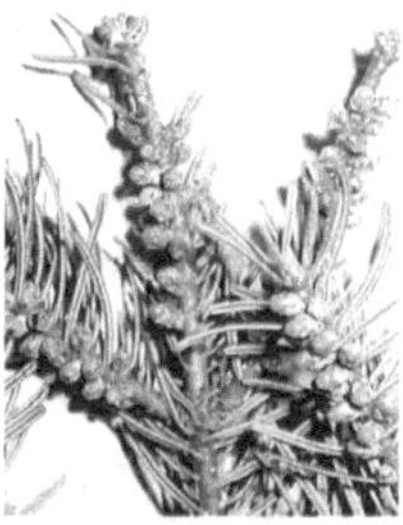
Male cones

Large Grand fir trunk

At Mount Rainier, Grand fir saplings grow along the banks of the Nisqually, White River, North Mowich, and Carbon rivers. Along the Nisqually, old trees exceed five feet in diameter.

Large individuals may represent remnants from a time when Grand fir inhabited terrain from the lowlands to the edge of the glaciers. This beautiful tree lives to an age of 300 years and may reach a diameter of four feet or more. Perhaps most easily identified by its long, lustrous needles, suitable habitat in the outwash environment along glacial rivers ensures Grand fir a continuing role in the forests of Mount Rainier.

Subalpine fir – *Abies lasiocarpa*

As the most common tree species in the higher elevation forests, Subalpine fir demonstrates an adaptation to a wide range of conditions. Here, trees must survive both the inundation of snowy winters and the stress of short, dry summers. Subalpine fir does so with a variable form, a capacity for sustaining a slow growth rate, and a versatile reproductive strategy.

Subalpine fir reflects the severity of local conditions in its configuration. Trimmed by the elements on desolate, wind-blasted ridges, the fir adopts a shrubby form no more than a foot or two in height and width. Nearby, in more benign conditions, a 50-foot tall tree scarcely resembles the ground-hugging shrub.

Stunted "krummholz" Subalpine fir

Short, stiff branches resistant to breakage from heavy snow jut out horizontally from the crown of tall Subalpine fir trees. Branches are retained throughout the length of the tree; often the lower limbs are much longer than the top ones. Brushing the ground and laying beneath the snow pack for several months each year, these branches may take root, producing clones. The parent tree and its identical offspring form tree islands, clumps that provide a mutually protective environment against the elements.

Female cone

Seeds are also a means of propagation. In late summer, two-to-four inch erect female cones may be seen on the topmost branches of mature trees. The quantity of the lightweight seeds varies within a two to four year cycle.

Subalpine fir seedlings are provided some protection against sub-freezing temperatures beneath a fluctuating snowpack. However, the snow also directly impacts the time available for photosynthesis. There is evidence that prolonged and warm summers have enabled establishment of this species in regions from which it was previously excluded.

Male cones

Major associates of Subalpine fir include Mountain hemlock and Alaska cedar. The distinctive needles are an aid to identification. They are about an inch long, bluish-green in color, with one white stomatal stripe on top and two below and brushed upwards from the twigs. Mountain hemlock bears its needles in a star like configuration and lacks stomatal lines while Alaska cedar is characterized by long sprays of scaly foliage. Another uncommon associate (at subalpine altitudes), Pacific silver fir resembles Subalpine fir, but its needles generally lack the top stomatal line, giving them a glossy appearance.

With a maximum diameter of two feet, in extreme environments the growth rate of subalpine fir may be less than an inch a year. A brittle trunk and thin bark make the fir susceptible to heart rot. This species is also short-lived (250-300 years).

Young trees bear thin gray bark which thickens and darkens with age, forming long, deep ridges.

Subalpine fir clump

The range of Subalpine fir extends from southern New Mexico to the Yukon in the north, east to the Rocky Mountains, and throughout Canada west of the continental divide. Within this wide territory, temperatures range from below freezing in the high mountains to desert highs, with equally variable precipitation. The presence of Subalpine fir throughout this region gives an indication of its adaptive capabilities. At Mount Rainier, Subalpine fir commingles with other species in the dense, middle altitude forests, but it is in the highest regime that the fir dominates. Here where the winds blow cool any time of year, the sturdy trees provide dark contrast to the high mountain meadows.

Noble fir -- *Abies procera*

Clinging to the steep hillside, the stout tree thrived in the sun above a forested canyon that stretched for miles below. Bracted, green cones perched at eye's level, decorating the tree top with their stout forms.

Cones, bark, branches, needles, configuration — all of these features facilitate tree identification. Of these, the female cones of Noble fir are perhaps the most distinctive. Longest of the true fir cones, they are stout, six inch-long structures with prominent bracts. Soft green during the spring and summer, in the fall, the cones mature to a purplish-brown. Like other true firs, the cones stand upright near the top of the tree where they disintegrate in place. The pollen cones are much smaller.

Noble fir needles

Mature Noble fir trees are typically free of branches for two-thirds or more of their length. The trunk beneath these limbs is straight and thick. Branches of mature trees are covered with stiff needles which point nearly vertically and bear two white stomatal stripes on the top and bottom. However, the needles of young trees bear little resemblance to the upturned needles of the older tree. Saplings bear narrow, soft needles; gray-green in color, they are arranged in a nearly horizontal configuration. A "hook" on the longer needles lays them at nearly right angles, while shorter needles line the tops of the twigs. This configuration resembles young Pacific silver fir, but the needles of Noble fir are more narrow. On occasion a sapling may be found with both the thin immature needles and the stiffer mature ones.

Large Noble trunk

The bark of mature trees is a distinguishing field mark. Gray and maroon in color, this beautiful feature is configured into long vertical plates, separated by horizontal lines which give the trunk a patchwork appearance. It is nearly two-toned, variegated in appearance, and, in good lighting, the plates have a gray sheen.

Noble fir is most commonly found in the middle altitude forest. A disturbance species with an intolerance for shade, small enclaves are typical, particularly on south-facing slopes. With its size and distinguishing bark, a single tree stands out from other species. Largest of the true firs, Noble fir achieves its great stature with a methodically slow but steady growth rate. Restricted to open areas by its shade intolerance, Noble fir may be at a disadvantage where its range overlaps that of Douglas-fir, another pioneering species characterized by a very rapid growth rate. In the middle altitude forest, however, a steady growth rate enables Noble fir to outdistance its associates, including the Douglas fir, which slows with maturity. Noble fir may reach a maximum height of 250 feet, and a diameter of over six feet, with a nearly ramrod straight trunk that reflects the steady growth rate. Near Tahoma Creek on the south side of the Park, the author measured a Noble fir nearly 21 feet in circumference.

Young, vigorous tree

With an altitudinal range of 3,500-5,500 feet, Noble fir grows in the region of deep and lingering snowpack. Common associates here include Western hemlock and Pacific silver fir. Soils are often shallow, but given sufficient moisture, the thin substrate is supportive, providing necessary drainage. Despite the snow, dry summers promote occasional fires, disturbances that provide suitable habitat for Noble fir.

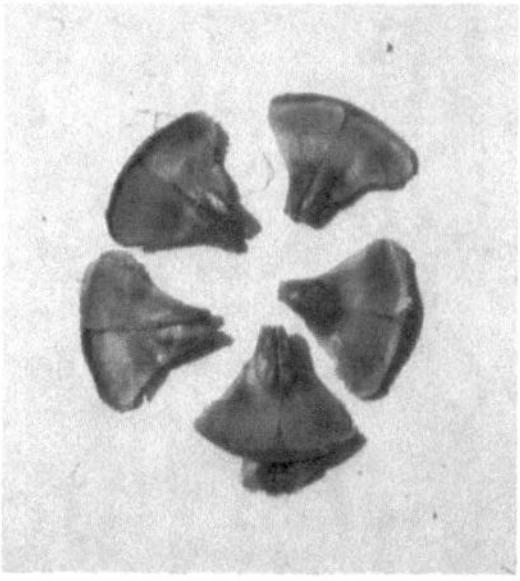

Large seeds

Resistant to disease and insects, with a maximum age of 600 to 700 years, Noble fir is a long-lived true fir. Although never common, Noble fir on occasion form nearly pure stands. Along mountainous highways or trails where the sloping terrain permits a tree top perspective, the stout cones of Noble fir may be seen in late summer. Dispersing large seeds in the fall, the cones seem a fitting touch on the tree that easily tops the canopy.

Alaska cedar -- *Chamaecyparis nootkatensis*

Draped in their yellow-green foliage sprays, the cedars hug the rocky, wet slope.

Draped in loose, swaying limbs, Alaska cedar is most commonly a species of wet habitats. In the Park, it grows from middle to subalpine elevations, often on north-facing slopes, or on rocky outcroppings near streams. Slow-growing, Alaska cedar may be confined to terrain where other species are excluded and often adopts a shrublike form. In this way, its strategy resembles Whitebark pine, a species of the loftiest habitats.

Alaska cedar limbs

Bristling with an array of chemical defenses against the attacks of insects and fungi, Alaska cedar may live 2,000 years. Thus, although restricted in range, Alaska cedar often outlasts its more common associates. In good conditions individuals may reach over 130 feet, with a tall, tapering trunk of over three feet in diameter. Deeply curved branches are covered in a soft foliage

Alaska cedar bark

with a slight prickle to the edges. The stringy, gray bark encases a warm yellow wood. Producing tiny seeds in half-inch cones which mature in two years, Alaska cedar also reproduces by layering. The most common coniferous species of high altitude avalanche chutes, this species persistently repopulates the fringes. Along fog-bound ridges whole thickets of curved limbs covered in yellowish-green sprays indicate the presence of a thriving community of relatives, while near rivers the gray stringy bark is an unmistakable field mark.

Female cone

With a thunderous clap, an avalanche bursts down steep chute, sweeping clean. Or so it seems. But in the spring when the sun melts the jumbled snow, bent forms pop loose from the icy bed. The flexible branches and yellowish-green sprays reveal an identity. Amongst the broken limbs of other species, Alaska cedar emerges as the lone tree survivor in the devastated landscape.

Male cones

Engelmann spruce -- *Picea engelmannii*

As if they wanted nothing to do with each other, the stiff needles spring upward and forward from the underlying twig.

The sixteen coniferous tree species at Mount Rainier are distributed among five genera. As representatives of the spruce genus, Sitka spruce and Engelmann spruce, occupy the entire altitudinal range of forested habitat. The larger Sitka is a lowland forest species, while the more common Engelmann grows in the middle and subalpine habitats.

The size and configuration of Engelmann spruce needles resemble that of Douglas-fir, but spruce needles are pointed and sharply tipped, a feature that may discourage browsing by herbivores. White stomatal stripes line the top and bottom of the blue-green needles; when they fall, brown stubs (pedicels) by which they were attached remain on the twig. Immature twigs are tannish and hairy, while new spring growth is a soft lime-green color.

Engelmann needles

Engelmann saplings are similar to Sitka spruce, but mature trees develop a tall, narrow crown with a slightly droopy appearance. Where they are free from crowding, individuals may remain limbed to the ground. Reproduction begins at a young age, although seeds are not produced in quantity until the tree is about 150 years old. In the late spring, the tops of mature Engelmann spruce are transformed with hanging male flowers and two-inch long reddish-purple female cones, all borne on the same tree. Mature female cones have thin, triangular shaped scales with wavy edges. As the season progresses, the female cones darken to a brown color, disperse their seeds, and sometimes drop, although the cones may remain attached to the tree for several seasons.

Cones on tree tops

Thin, reddish-purple bark cloaks a

straight trunk which may stretch to 80 feet at timberline, and to 140 feet at lower elevations. An individual tree along the Summerland trail measured fourteen feet in circumference. The distinctive bark is broken into small, platy patches which leave pale scars when they fall. Englemann spruce may live to 400 or 500 years, a "middle" age for a coniferous species.

Engelmann trunk

Outside of the Park Engelmann spruce grows throughout an extensive range. Established farther north than any other conifer except one (likewise a spruce), its domain also extends to the south and east. Engelmann spruce thrives in cool sites and is more commonly found east of the Cascade crest than on the western slopes. At Mount Rainier, this species is confined to the north and east side of the Park where large trees are found on high elevation slopes, glacial moraines, and in wet meadows.

The restricted range of Engelmann spruce at Mount Rainier suggests a couple of possible origins for this species. The population may be a remnant of a forest that encompassed much larger areas during the ice ages. Possibly the spruce has pioneered the high altitude slopes by a westward movement over the crest.

Highest elevation Engelmann spruce

Shrublike individuals hug rocky slopes at over 7,000 feet of elevation near Sunrise, possibly the highest elevation for any tree species. Extending its range into the lower reaches of the White River valley the versatile Engelmann grows in close proximity to Sitka spruce 4,000 feet below.

Engelmann spruce commingles with several coniferous species, including Pacific silver fir, Alaska cedar, Subalpine fir, and on occasion, Whitebark pine. At the higher elevations, this species grows in association with the firs on warm southern slopes and in meadows, while in the lower regions, it shares cool, moist habitats along sheltered streams with Alaska cedar. Englemann spruce also does well on recently deglaciated sites, such as that found at the snout of the Winthrop Glacier.

Burroughs Mountain — Engelmann spruce habitat

In the late summer, when the meadows are emblazoned with flowers and the last snows have disappeared, you may be lucky enough to notice the cone-covered tops of Engelmann spruce. Maroon-colored from hundreds of cones, the sturdy spires tower above their associates, where they sometimes invite destruction in a storm-induced lightning strike. Growing in stunted form high on the alpine slopes, Englemann spruce survives in a habitat shared by diminished alpine plants.

Sitka spruce -- *Picea sitchensis*

December slides into January, and glowering skies scrape the earth, obliterating the difference between day and night. The inundation has grown tiresome--it is winter in the Northwest, a time when gratitude for rain fades as quickly as memories of summer.

Near the Pacific Ocean, annual rainfall often exceeds a hundred inches. The effects of such abundance is evident in Sitka spruce, one of the largest conifers on earth. Cloaked with limbs which may sweep outwards 30 feet or more, the massive Sitka can reach a diameter of over sixteen feet in this salty, wet habitat.

Although Sitka spruce attains its greatest size here, this species is not restricted to the narrow coastal strip. Far from the ocean's roar, Sitka spruce plays both a pioneering and climax role in low elevation wet sites at Mount Rainier in the Carbon and White River valleys. Formed of glacial melt waters, the these rivers transverse habitats ranging from subalpine to lowland wet environments reminiscent of coastal environments. Wetland forests near the lower Carbon River include several old members of this species, some greater than four feet in diameter. The author has measured one individual with a circumference of nearly eighteen feet. Perhaps more surprisingly, because of its cold temperatures and distance from the coast, the White River valley also supports a small forest dominated by this species. In this region, Sitka spruce seedlings are interspersed with mature trees where the river has cut a wide, densely forested valley. At an elevation close to 3,000 feet, the Sitka range nearly overlaps with that of Engelmann spruce, a subalpine species. The spruce may also be found in a marshy habitat along the Nisqually River near the Park entrance.

Giant limbs

Sharp-tipped needles

Sitka spruce may achieve a phenomenal growth rate, which often continues into middle and old age. The otherwise straight trunks are occasionally distorted by large burls near their base, giving a grotesque appearance to the stately trees. Although the light requirements are greater than another lowland species, Western hemlock, the spruce may compensate for this requirement with the fast growth rate, particularly in beneficial habitat. This capability enables the tree to literally rise above other species which may be more shade tolerant, allowing it to reach an incredible 300 feet in height.

Bark of old Sitka

Bluish-green, inch-and-a-half long needles encircle light brown twigs. Sharply-tipped needles provide an aid to identification The best field mark however, may be the bark--only the Sitka's relative, the Engelmann spruce, has a similar covering. Maroon-colored in mature trees, the thin bark is broken into rectangular fragments which resemble puzzle pieces. This unusual bark provides little protection against fire, but in the wet habitats where Sitka spruce grows, fires are more easily suppressed than in drier regimes. However, during the summer, prolonged drought in the Northwest may lead to fires in the tinder-dry landscape, and forests that are home to the Sitka spruce may be decimated.

Old female cones

Early season female cone

Light brown cones range in size from two-and-a-half to four inches and are covered with thin, wavy scales. Following dispersal of the tiny seeds by the coastal winds, the cones drop intact to the ground from their position high in the tree's crown.

With its long branches, patchwork bark, and massive trunk, Sitka spruce resembles no other tree. Although it lacks the stateliness of the firs or the exotic beauty of the cedars, for sheer size and versatility this species is arguably unmatched in the Northwest forest. In this land of salty rains and winter snows, the spruce comes into its own, thriving in coastal and mountainous environments alike.

Male cones

Whitebark pine -- *Pinus albicaulis*

A thousand feet above the crowded forest, a lone tree with stretched and arched branches hugged the ridge line.

Whitebark pine grows near the upper limit of plant life at the mountain. Not only does this species inhabit the loftiest regions, but it does so exclusively, rarely growing below an altitude of 5000 feet, and restricted to the north and east sides of the Park.

At the lower elevations, where it grows in association with Mountain hemlock, Subalpine fir, and Alaska cedar, Whitebark pine adopts a conventional tree form, with a single trunk topped by a spreading crown of stiff branches, sometimes 40 feet above the ground. At higher elevations, the pine often develops a base several feet in diameter composed of multiple trunks. Where exposed to wind and severe weather, Whitebark pine may resemble a severely pruned shrub. Age is difficult to determine in these contorted individuals; here a shrub a few feet high may be 500 years old.

High elevation tree

Large tree at Sunrise

Bunches of five yellowish-green stiff needles three inches in length distinguish this species from Western white pine, which bears longer, more flexible needles. The smooth and whitish bark of young develops into a scaly, crumpled-looking gray covering with age. Three-inch long purplish cones up are egg-shaped; requiring two years to reach maturity, they remain on the tree for several years before disintegrating. The large seeds,

Male cones

Needles

eaten or stashed by birds, may pass through the host's digestive track unaltered; both mechanisms may serve to disseminate the seeds far from the parent tree.

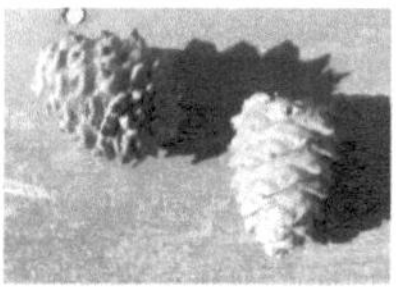

Female cones

Whitebark pine embodies hardiness. Assuming a prostrate form where necessary, this species adapts equally well to a more benevolent environment where it provides food and shelter for a community of plants and animals.

Lodgepole pine -- *Pinus contorta*

Possessing neither the grandeur of a Douglas-fir nor the majesty of a Western white pine, Lodgepole pine rarely inspires accolades.

With a maximum diameter of two feet and a life expectancy of less than 250 years, in terms of size and longevity, Lodgepole pine is a small conifer. As a lesser component of the western Cascades forests, this pine is found in small groups and in dense stands, where it grows to the near exclusion of other species.

Although Lodgepole pine may occupy areas inhospitable to other species, this opportunist also invades disturbed sites where competition is keen. Producing and dispersing millions of tiny seeds may enable this pine to compete with faster growing species adapted to the openness characteristic of these locations.

Female cone

Hanging inside their cone cocoons, Lodgepole seeds remain viable for years, frequently long enough to become entombed within the cone shell in the tree's bark. Cones often produce a resin which seals the seeds, enabling them to survive a conflagration while, at the same time, requiring fire for release. This strategy may leave the seeds intact on the ground or attached to the par-

Male cones

Needles

ent tree until sufficient heat melts the resin. Dispersing quickly, a dense Lodgepole stand may develop before other species invade.

The food and water demands of a stand of young trees is ultimately limiting. Rapidly absorbing nutrients, the trees may now appear sickly, their longevity jeopardized. Subsequent natural thinning sets the stage for an invasion of other species.

Needle bundles composed of two one-to-three inch needles are unique amongst the pines; other pine species at the Park bear needles in groups of three or five. Dark, thin bark becomes roughly furrowed with age. Shallowly rooted, Lodgepole pine tolerates a range of soils, including chemically-laden hot springs.

Old trunk

Lodgepoles are intolerant of shade, but can survive the cold, and are found in protected, frosty areas. This pine is an important component in the forests at Mount Rainier. Rarely dominant, it may be found on burned sites, rocky habitats, or other disturbed areas. Pure stands occur in high elevation habitats, on recently de-glaciated terrain, and in fired middle altitude sites. At these inhospitable locations, this species enhances the site for plants and animals alike.

Dense stand

Western white pine -- *Pinus monticola*

Rarely represented by more than a few individuals at any site, as with Lodgepole pine, Western white pine is a disturbance species. Unlike its smaller relative, however, Western white pine sustains a potentially vigorous growth rate, often exceeding a foot each year. A species of the middle altitude and subalpine forests, Western white pine often grows in association with other pioneers, such as Douglas-fir. At other locations throughout the west, this species may form pure stands over a range of elevations, including lowland sites along the Northwest coast.

In the middle altitude habitat at Mount Rainier, the growth rate of Western white pine sometimes surpasses Douglas-fir. Seasonal growth is easy to determine, as the pine produces a new whorl of branches near the top of the tree with each cycle. The branches appear to stretch upwards, like giant arms, in an open configuration.

On mature trees, distinctive, checkered gray bark covers a nearly cylindrical trunk. Tending to drop the lower branches, four-foot diameter individuals 150 feet in height tower over many associates. In the Ohanapecosh River drainage a tree over fifteen in circumference was measured by the author.

Vigorous growth and cones

Borne five in a bunch, the soft, blue-green needles are lined with white bands on their bottom surface and are two-to-four inches long. Nine-inch long cones, the largest of any Park species, hang near the tops of the trees. Requiring two years to mature, the cones bear small seeds that are nonetheless relished by squirrels whose caches may play an important role in seed dispersal. Wind also disseminates the seeds.

Needles

This easily identified tree is among the most beautiful of the pines. A large pine cone on the forest floor gives away its presence, as do the tall open configuration and the luxuriant blue-green needles. Growing vigorously on dry and thin soils, Western white pine may attain great size in such impoverished conditions. Although it is never common, small communities are occasionally found on rocky outcroppings, and individuals are present throughout the middle altitude forest at Mount Rainier.

Male cones

Ponderosa pine -- *Pinus ponderosa*

In the Northwest, Ponderosa pine is most commonly found east of the Cascade Mountains' crest. Here this magnificent pine grows near the hot and dry edge of the desert floor where, in regions of ample moisture, 200 foot trees may attain a diameter of nine feet. An adaptation to arid habitat is evidenced throughout this species range, which includes the higher elevations of the Sierras, the Rocky Mountains, and the mountains of Idaho and eastern Oregon.

Given this preferred habitat, the wet, cold environment at Mount Rainier seems ill-suited for Ponderosa pine. And yet the species is here, reduced in stature, but present in small numbers on the Park's east side, closest to the Cascade crest. Very scarce, the method of establishment for these few is difficult to determine. Perhaps the seeds were carried to these sites by animals.

Ponderosa pine is readily identified by needles borne in bunches of three. The scaly, dark gray bark of young trees becomes reddish and heavily grooved with age.

Small Ponderosa — east side

Ponderosa pine may bear its cones as early as seven years and continue to produce well beyond a prime age of 300 years. The cones are egg-shaped, to five inches long, and adorned with a sharp prickle on each scale. Ponderosa pine grows on a variety of soils and may live to 500 years; in the past, old forests dominated by this species covered vast tracts of land.

Although Ponderosa pine inhabits dry

Deeply grooved bark

regions, its growth is nevertheless restricted by available moisture. Typically requiring at least ten inches of rain each year, this species may send long tap roots into the soil, sometimes deeper than the height of the tree. Intolerant of shade, prior to an instituted policy of fire suppression, thick-barked old Ponderosa pine trees were fire resistant. Old pines are often marked with a black slash, an indicator of a fiery battle fought and won. Ponderosa pine will not replace itself in areas where it must compete with more shade-tolerant species.

Two-year cone

The life span of the few Ponderosas at Mount Rainier may be limited by the unsuitable conditions. If you photographed a Ponderosa pine tree today and returned to the same location the following year, you might be surprised to discover that the individual had lost the struggle, its drying needles or prostrate form indicating defeat. On the other hand, climatic changes or local variations in forest composition may promote a continued presence for this rarest of trees at the mountain.

First-year cone

Douglas-fir -- *Pseudotsuga menziesii*

With intractable origins in another land, it embodies a primitive idea and has stood the test of time. Long-lived, a survivor of fire and wind, it has unwittingly persevered over upstarts. The ultimate conifer, it is an unparalleled botanical success story.

A fir which isn't a fir, and a cosmopolitan species with few living relatives, Douglas-fir evolved far from its Northwest home. The genus name *Pseudotsuga,* meaning "false hemlock," reflects the confusion amongst the early European botanists about correct family

Douglas-fir bark

membership for this ubiquitous tree. First identified as a pine, then as a true fir *(Abies),* finally, after discovering a relationship to species found in China and Japan, the *Pseudotsuga* genus was settled upon. Much of the confusion undoubtedly came from the fact that different parts of Douglas-fir do resemble other species. For example, the blistered bark of young trees is similar to the true firs. But Douglas-fir has only one relative in the Western Hemisphere; this is the much smaller Bigcone Douglas-fir, a southwestern species.

Douglas-fir is a pioneering, sun-loving tree on sites ranging from zero to over 4,500 feet. With a growth rate rarely matched by other species, if success is measured by numbers, on lowland disturbance sites no other species matches Douglas-fir. It is an enormously successful species.

Fire-scarred Douglas-fir

Old trees may live a thousand years, a longevity aided by a thick bark. Providing a defense against the constant threat of forest fire in an environment of dry summers and moderate winters, the bark is a definitive field mark in mature trees. Deeply furrowed, with alternating brown and tan stripes, the bark may reach a foot in thickness; no other species is protected so thickly or distinctively. Long reddish cracks mark the beginnings of the vertical grooves and are not uncommon in trees less than a foot in diameter. The bark of old trees is often blackened with the scars of fire. These survivors make clear the advantages of allocating resources into the thick bark.

Female cones

With their pale gray bark, small trees may be distinguished from other similarly clothed species by the pointed, scaled new branch tips. The needle configuration also provides a unique field mark. Encircling the twigs, the needles are medium green in color with two white lines on the underneath side. In the spring, the tree is brightened by the lime green color of new growth. The foliage has an unmistakable evergreen scent. Soft three-or-four inch cones are a further aid to identification. Borne in profusion on the branch tips throughout the tree's length, they litter the ground at the season's end, each decorated with thin protruding bracts adjacent to individual scales.

Male cones

The quick growth rate of Douglas fir is manifested in a loose, sweeping configuration. Old trees typically support an irregularly formed, wide crown, an excellent field mark where the trees tower above the forest canopy. Douglas fir saplings are common in lowland disturbance sites; these include the edges of lowland roads, stream banks, de-glaciated terrain, mud-altered, and burn sites. The sunny quality of these sites suits the fir, but with the heightening of the canopy, this species is gradually replaced by more shade tolerant trees, such as Western hemlock. Only the oldest trees survive tenaciously above the more shade tolerant species.

Douglas-fir crown

Thus, the number of Douglas fir trees gradually diminishes measured from the time since disturbance. The oldest sites, sometimes burned more than once, usually are distinguished by a few gigantic members, the lone survivors of the conflagration. At the Park, where Douglas fir is a common lowland species, these old trees, some more than eight feet in diameter, populate forests characterized by varying densities of trees, herbs, and shrubs. Where long shafts of filtered sunlight strike the forest floor, a rich understory flourishes. The variety seems endless here, but the communities of these ancient sites provide a window to a past where immensity and diversity were commonplace.

Western redcedar — *Thuja plicata*

Wrapped in strips of clean bark, a buttressed trunk narrows to a tapered top.

Western redcedar requires abundant moisture for germination, a restriction that promotes establishment along stream banks or in wetlands. Most common below 3,500 feet, the range of Western redcedar extends to the middle altitude forest. There it grows on steep hillsides as well as stream valleys where soils may be very thin. At higher elevations this species is increasingly replaced by the related (although of a different genus) Alaska cedar.

Western redcedar is not a cedar but a cypress. Fernlike leaves arranged in yellow-green sprays draped from down-curved branches distinguish this conifer from all others except Alaska cedar. The stringy, reddish bark may be pulled lose in long strips, and was utilized by Native Americans for a variety of purposes. The bark resembles that of Alaska cedar, but the latter has a gray bark that shreds less easily.

Fernlike foliage

Western redcedar branches impart a sweeping configuration to the trees, and the heavily buttressed trunk tapers quickly. Half-inch light brown oval cones produce 15 to 20 seeds each; Western redcedar is second only to Western hemlock in seed production.

Old trunk

Most conifers provide only minimal provisions in their seeds, making the seedling highly dependent on the environment for sustenance and growth. Western redcedar carries this scarcity theme to an extreme; it takes 400,000 seeds to weigh a single pound. Many of these minuscule seeds germinate, but most seedlings never mature. Succumbing to drought, buried by forest litter, or con-

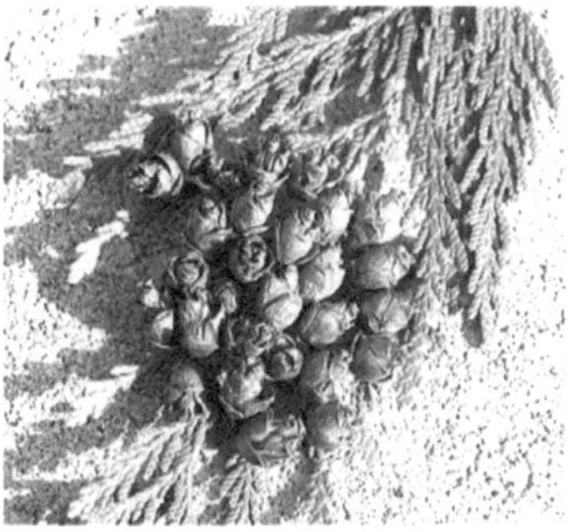

Old female cones

sumed by animals, they are not wasted. Those seedlings that do survive may grow to truly gigantic proportions, on occasion achieving a circumference of sixty feet. Western redcedar also commonly sprouts from rotting trunks and fallen branches to form nearly exclusive communities; these individuals may have a higher success rate.

Once established, Western redcedar may live to more than 1,000 years, commonly reaching seven or eight feet in diameter, and 200 feet in height. This immensity may be partially attributed to the wet, fertile habitats where it grows, but of equal importance to longevity is near immunity to attack by insect and disease. Few invasive fungi common are able to survive the highly acidic bark of this species. The bark remains clean, giving a new appearance to old trees. Insects are thwarted by the cedar's chemical defenses, a strategy that may exclude other species from the immediate vicinity.

Male cones

In spite of a shallow root system, Western redcedar is resistant to windthrow, and although the thin bark makes this species vulnerable to fire, the preference for wet habitats reduces the threat.

Second only to Western hemlock in shade tolerance, Western redcedar fills both climax and pioneering roles in the forest. Rarely forming large stands, in the absence of disturbance, Western redcedar may achieve a temporary dominance, outliving its associates. In the quiet of a lowland marsh, Western redcedar thrives unobtrusively where delicate needles provide striking contrast to its massive trunk.

Western hemlock -- *Tsuga heterophylla*

Beneath the towering firs, seedlings of another species extend their lacy fronds into pools of filtered light. In silent lines on fallen giants, they grow slowly, unhurried inheritors.

Western hemlock fills the "climax role" in lowland Northwest forests, meaning that this species replaces itself and other species as well. Thus, in aging forests Western hemlock of many ages dominate the community. Shade tolerance and an ability to grow slowly beneath the canopy ensures its position. As other species die out, young trees may experience an accelerated growth rate in openings created by fallen trees or other small disturbances.

Western hemlock employs an effective strategy for capturing sunlight in the shady forest. Each twig is covered with densely packed needles of varying length, thus exposing a maximum surface area to the filtered rays. Short and soft, the needles impart a delicate appearance to the tree.

Multiple-length needles

Western hemlock often plays a minor pioneering role. In the company of species characterized by faster growth rates, longer life, and greater size, Western hemlock grows at a moderate rate. As the forest ages, the increasingly shady environment on the forest floor excludes pioneers such as Douglas-fir, permitting Western hemlock to dominate.

Male cone

In the lower elevation forests at Mount Rainier, Western hemlock seedlings grow on fallen logs common in this habitat. Often densely packed on these elevated platforms, the seedlings may absorb sunlight from above and nutrients from below.

Western hemlock is easily identified by its needles, cones, and droopy top (leader) which may hang loosely as much as 200 feet above the forest floor. Western hemlock may grow to four feet in diameter, with a longevity of 500 years. Near Tahoma Creek on the south side of the Park, an old Western hemlock measured by the author measured nearly six feet in diameter). All of this begins with a tiny seed encapsulated within inch-long cones. Borne on the previous year's branch tips, thousands of these cones fall each year, littering the forest floor, and revealing the presence of a mature tree.

New female cone

Old trunk

When compared by size and longevity to Douglas-fir, Western hemlock may give the impression of inferiority. But the mature hemlock embodies a role that to a great extent defines the old growth forest. A common species from sea level to over 4,000 feet of elevation, with its sagging top, soft irregular foliage, and tiny cones, Western hemlock is an easily identified conifer.

Mountain hemlock -- *Tsuga mertensiana*

By contrast to the symmetrical spires of Subalpine fir, the most common tree species of the higher elevations, Mountain hemlock appears loose and unkempt. Arranged in star-like bunches, the densely packed bluish-green needles are brightened by white stomatal lines on both surfaces. They cover branches of multiple lengths, imparting an irregular geometry that reveals a similarity to a related species of lower elevations, the Western hemlock.

The gray to brown bark of Mountain hemlock becomes thick and furrowed with age, an adaptation that seems an unnecessary protection in wet habitats. Possibly the bark reflects an evolution under different conditions, or offers added protection against fungi and insects. Slender one-and-a-half to two-inch purplish cones are borne annually; their color becomes tan with maturity.

The sweeping branches slough off the heavy snows characteristic of the subalpine environment. In sheltered sites Mountain hemlock grows

Male cones

Needles

slowly to a height of 50 feet or more, while in more severe locations it assumes the form of a contorted shrub. Inundation with up to 30 feet of snow severely limits growth opportunities, shortening photosynthetic opportunity. Mountain hemlock also grows in middle altitude forests, where it may reach a height of 100 feet with a diameter of three or four feet. Here the needle configuration helps to distinguish this species from Western hemlock.

Female cones

Thick bark on old tree

Often the largest of the subalpine species, individuals appear a bit out of place, as if they descended from an escaped ancestor adapted to a wet, warmer landscape. Their presence is nonetheless beneficial to a community of plants and animals that find a protective environment near the great trees.

Deciduous species

Vine maple -- *Acer circinatum*

The bright green leaves of summer days become the fiery reds and oranges of autumn. Star-like, they appear frilly beneath gigantic Western hemlock and Douglas-fir.

A species of the low to middle altitude forests at the mountain, with its seven-to-nine pointed starry leaves, Vine maple is easy to identify. In the lowlands, this species forms small leggy clumps beneath the towering conifers, and is one of the few deciduous species of its size in this understory role. In the middle elevations Vine maple often dominates avalanche chutes. Forming pure shrubby stands in these steep areas, vine maple bends rather than breaks beneath sliding snow.

Star-shaped leaves

Small, reddish flowers

Whether under the canopy or in an avalanche chute, Vine maple is shrublike in appearance, often as wide as it is tall. On the slopes, it may be distinguished from Slide alder (the other common species of these regions) by leaf shape, the bright green leaf color, and the slender red twigs.

Vine maple bears small reddish flowers in early spring; these mature to double-winged seeds characteristic of the maple family. The star-shaped leaves have seven to nine points and are borne on flexible red stems. Delicate and lacy, Vine maple nevertheless is a rugged species, thriving on steep, rocky slopes. Laying its many branches horizontally, the beauty of Vine maple may become a curse to the hiker trying to negotiate a climb through snarled limbs!

Autumn hillside trees

Although unobtrusive beneath the forest canopy, in autumn the brilliant orange and red leaves of Vine maple add a splash of color to forest floor and avalanche chutes alike. Then the conifers and gray rocks seem to recede beneath the red glow of this hardy species.

Seeds

Bigleaf maple
Acer macrophyllum

Giant green leaves hang indolently from curved maroon twigs. Beneath arching limbs a park is sculpted, an inviting respite from the summer sun.

During the summer it is unlikely that the 8-to 12-inch long leaves of Bigleaf maple could be confused with any other species. Bright green, the leaves are deeply cut deeply into five lobes. After the leaves fall, the trees are most recognizable by their thick trunks, often decorated with licorice ferns, or covered with moss. Old trees support spreading crowns of massive limbs that create a tree as wide as it is tall.

Leaf

At Mount Rainier, Bigleaf maple is a lowland species with a preference for alluvial soils along glacial streams and rivers. In these wet habitats, common deciduous associates include Black cottonwood and Red alder. Where moisture requirements are met, Bigleaf maple also pioneers sunny slopes, where it grows on a variety of soils.

Mossy trunks

Shallow-rooted, Bigleaf maple prefers open, sunny locations, although it will tolerate some shade. Intolerant of flooding, this species nonetheless grows best in those areas that are subject to occasional inundation. Individual trees may live to 200 hundred years on favorable sites; rarely forming pure stands, Bigleaf maple is opportunistic, thriving as part of the seral community. Here it grows vigorously in association with other deciduous species, and conifers such as

Young maple trunk

Douglas-fir and Grand fir.

In the spring, long clusters of fragrant yellow flowers offer succulent nectar and pollen to numerous honeybees and insects. This lure serves a vital function, for Bigleaf maple is dependent upon the insects for fertilization. The flowers mature into fruits (samaras), a doubled nutlike seed shaped like a swept wing. The seeds germinate the following spring, but only a few will survive. This species also reproduces vegetatively, sprouting in large numbers from old stumps.

Flowers

Samaras

The large leaves turn golden in autumn. In death, the leaves serve the community by enriching the thin and unstable soils with their organic content.

With its leafy show, Bigleaf maple imparts a luxurious beauty to the landscape. Along the streams and rivers at Mount Rainier, the giant trees with their immense eaves seem to provide a link between the mountains and the coastal lowlands miles away.

Red alder -- *Alnus rubra*

Dangling in anticipation, male red alder catkins add a warm color to the cold stream banks. The dormant conifers give little indication that spring is near.

As the most common deciduous species in the Park, Red alder thrives along stream banks and other lowland sites. Following a disturbance, such as a flood in which the banks are inundated or swept away, Red alder is typically the first species to reinvade the site. On recently de-glaciated land or in fire-swept terrain, Red alder fills a pioneering role. Whole forests of alder grow quickly, but, lacking further disturbance, they are replaced with coniferous species, such as Western hemlock or Western redcedar.

Old bark

Thin gray bark lightened by splotches of lichen covers trunks and branches. Sometimes reaching nearly to the ground, the branches bear light grayish-green oval leaves with serrated, single teeth. Four-to-six inches long, the leaves turn to a muted brown in late fall when dark cones decorate the trees; these female structures are not cones at all but woody catkins holding within the tiny seeds. In the spring, male catkins hang from bare branches; upon shedding their pollen, they quickly disintegrate. The alders may reach 100 feet in height, and form dense communities of leafy crowns and curved trunks.

New bark

Red alder rarely receives the accolades showered on a common associate, the gigantic Douglas-fir, as it is neither beautiful nor visually outstanding. Such a judgment underestimates the importance of the alder, however, as it is this species that employs an uncommon biological mechanism for capturing nitrogen, an element vital to forest health. Plants utilize nitrate, a nitrogen and oxygen compound formed from a series of chemical reactions. Most trees cannot acquire nitrogen directly from the atmosphere or the soil; the ensuing nitrogen deficiency may result in a slow growth rate. Red alder absorbs atmospheric nitrogen directly into its leaves and transports this element to root nodules where it is made available for reactions within the soil. When they fall, the leaves enhance the soil directly with this vital element.

Leaves

New female cones

As a pioneering species Red alder is intolerant of shade. It sprouts and shoots up quickly, sometimes achieving heights of 15 feet in five years and 80 feet in 25 years, and is most noticeable as the available space be-

Male catkins

comes crowded. At the same time, the alder is placed in a precarious existence by the dynamic nature of the streamside environment. If it survives to maturity, the growth rate slows; this species rarely lives longer than 100 years.

On sufficiently stable sites, more shade tolerant species grow within the aging alder community. These replacements are the foundation for the climax forest. However, a continuing role for Red alder is insured by an environment where the forests are subject to frequent disturbance. With its nitrogen fixing mechanism, this unobtrusive species brightens a cold winter scene, while playing an important role in the vigor and diversity of the forests at Mount Rainier.

Slide alder -- *Alnus viridis* ssp. *sinuata*

Dying glaciers fall back, leaving barren soil peppered with foreign rocks. A tantalizing invitation to life, the landscape promises little to sustain it.

Equipped for survival in the rockiest terrain, opportunistic plants press against glacial boundaries. Among the most rugged of these is Slide alder, a common pioneer on sterile glacial soil. Similar to the related Red alder, Slide alder absorbs and stores atmospheric nitrogen; by transporting this vital element to the sparse soil, the alder and succeeding species are provided a boost for germination. This capability may spell the ultimate demise of Slide alder, for, as the soils are enriched, this species loses the competition for living space to other species.

Nevertheless, in landscapes where change is the rule rather than the exception, Slide alder is ensured a continuing role. This species is well adapted to avalanche-prone terrain. As the snow pack deepens, alternating thawing and freezing cycles destabilize the accumulated mass, and, loosened by the pull of gravity, the sliding snow strips the terrain. When the ice melts, exposing the sterile rocks, the landscape may be void of all vegetation except the battered forms of Slide alder.

Leaves and new male cones

Male catkins

This unobtrusive plant has taken an adaptive form that aids survival in the unstable environment--shrublike, its flexible branches bend beneath the onslaught.

Slide alder bears greenish-yellow, ovate leaves. They are doubly toothed, up to five inches long, and ribbed with straight parallel veins. Light gray bark is often covered with lichen. During the spring, long male catkins composed of tiny yellow flowers hang from bare branches; the female flowers resemble small reddish cones. When fertilized they mature to a hardened, brown structure, each containing tiny winged seeds. Slide alder usually grows above an altitude of 3,000 feet, but it also inhabits moist canyons where avalanche chutes extend to lower elevations. At these sites, the alder may take a conventional treelike form. The more typical horizontal and shrublike configuration is a headache to any hiker attempting to traverse alder-occupied terrain. Bending but not breaking and difficult to negotiate, the frustratingly adaptive strategy of Slide alder may be experienced firsthand.

New female cones

Admiration may be easier from a distance. If you see a leggy shrub on a steep hillside or a spot of green on a glacial moraine, there is a good chance that it is Slide alder. Nestled against the glacier's edge, turning pale green in the spring, the alders provide a light contrast to the dark sea of surrounding conifers. Residents in a barren land, these blind pioneers pave the way for other species.

Black cottonwood -- *Populus balsamifera* ssp. *trichocarpa*

Along lowland rivers the crowns of Black cottonwood tower over smaller trees. Easily overtopping Bigleaf maple, the second largest deciduous species at the Park, large triangular cottonwood leaves transform solar energy into food for the entire tree. The crown serves more than one function, for within is provided a home and haven for many animals, including large raptors, such as the bald eagle and osprey.

The size of Black cottonwood is not acquired without a price. This species requires copious amounts of water and is most often found in a streamside habitat. The dark trunk becomes deeply ridged with age and holds water reserves which are readily released from any wound. If moisture requirements are met, Black cottonwood will inhabit otherwise dry regions, and thus is found along rivers east of the Cascades. Where sufficient moisture is available, individuals also become established on steep slopes in middle and high elevations.

Triangular leaves

A member of the willow family, Black cottonwood bears male and female flowers on separate trees. Following fertilization, the female flower develops into a grapelike cluster which releases plumed seeds late in the summer. These seeds, which give the cottonwood its name, drift in the breeze, filling the air with a cloud of white. Black cottonwood also regenerates by cloning, and it is not unusual to find stumps supporting a vigorous community of sprouting limbs.

Massive trunk

In autumn, the two-toned green-and-white leaves turn yellow, and Black cottonwood wears a golden crown. During the winter, the massive and leafless top stretches skyward, where it is subject to breakage from the accumulation of snow or ice. With the passage of a winter storm, the forest floor may be littered with the shattered branches.

In the wettest habitats, Black cottonwood may reach a height of 175 feet and a diameter of five feet. Young, yellowish-gray bark turns black and furrowed with advancing age, and death comes at a maximum age of 200 years. Old rotting stumps have their use, providing habitat for birds and mammals in their decaying bulk. Even seedlings bear full-sized leaves on their branches. Along the rivers and streams, the majestic Black cottonwood forms a bright backdrop for the smaller deciduous species and conifers. Black cottonwood puts on a colorful show Large trees accent the winter scene with their distinctive angularity. And, like the giant Northwest conifers, Black cottonwood achieves a size unmatched by any other.

Male catkins

Grooved bark

Bitter cherry -- Prunus emarginata

An uncommon species, Bitter cherry is a pioneer of moist, open lowland habitats. Although its growth rate does not match other disturbance species, such as Red alder, which typically overtops it, the cherry nonetheless grows rapidly. Its soft wood is prone to heart rot, however, so that in less than half a century this short-lived species succumbs and falls. Following the decay of the heartwood, it is not unusual to find the more durable red bark left as a skeletal case on the forest floor.

A member of the rose family, Bitter cherry is most easily identified during the spring when its small, notched, five-petaled blossoms create bright spots along lowland roads and wetland borders. Yellow stamens protrude from the flowers, attracting insects. Other distinctive field marks include the bulbous protrusions at the base of the leaves.

Old trunk

Bitter cherry's name may be derived from either the acrid fruits or the acidic bark. Unpalatable to humans, the half-inch long ripe cherries are consumed by rodents and birds. The large seeds pass unscathed through the animals' digestive track, and are deposited wherever the carrier goes. This effective dispersal mechanism provides the cherry with a multitude of possible germination sites; thus, the short life span may not be such a disadvantage for a species which can so easily disperse its seeds.

Blossoms and leaves

Bark with lenticels

The smooth, chestnut-brown bark is covered with light horizontal bands (lenticels) which allow air to enter the tree. Bitter cherry may reach a height of 50 feet and a diameter of 12 inches during its brief life. The trunk grows firm and wide, and the branches form a loose oval crown which extends for much of the length of the trunk. Toothed, one-to-three inch pale green leaves, covered with down on the underside, bear two distinctive small glands at their base. In autumn, the leaves are transformed to a muted brown, and, falling, leave the fruit tree's limbs bare for the winter season. Then, this species blends well into the somber forest, where it awaits the warm days of spring.

Willows
Salix scouleri
Salix lucida
Salix sitchensis

Scouler's willow

If learning to distinguish the conifers seems daunting, tackling the willows is unquestionably an exercise in futility! There are over 300 species on earth, 30 in the Northwest alone, and at least three tree-sized willows at Mount Rainier. To complicate matters, they occupy similar habitats, closely resemble one another, and, as if to be contrary, hybridize with apparent abandon. However, willows can be distinguished from other deciduous trees quite easily, and fortunately, the number of species at the mountain is small.

Sitka willow catkins

Perhaps more important than being able to separate the willows is the recognition of the important role that these unobtrusive trees play in the forest community. Willows produce the tiniest of seeds, each attached to a little tuft which is easily dispersed by winds. Too small to provide sufficient food for the seedling, the seeds must germinate quickly in a restricted environment, usually in continuously moist soils. At the mountain, the most consistently wet habitats are found in sandy banks along streams and rivers. Willows often grow in densely packed communities in this environment where they help stabilize the banks, sometimes thriving where the associated Red alder cannot thrive.

Pacific willow stipules

During the winter, willows brighten an otherwise gray landscape with twigs of yellow, red, and green. In the early spring, trees bearing the female and male catkins are borne on separate trees. The male catkins resemble the female blooms except for the profusion of yellow stamens which reach extend beyond the flower stalk. At this time of year when food is scarce, willows offer nourishing pollen and nectar to insects, which, in payment, provide a means of dispersal. Small birds find good nesting habitats within the protective willow branches.

Willow leaves are usually longer than wide, and are attached by short stemmed branches which have small earlike structures at their base. These stipules, along with the distinctive buds, which are enclosed in a scale, provide excellent field marks. The various species often resemble shrubs more than trees, sending up multiple stems, creating dense thickets.

Scouler's willow

As with other deciduous species in the Park, willows are often found in the lower elevations, but they will also grow in the subalpine environment. Here they crowd and alter the wetland habitats, and may provide a small riparian zone along sparsely populated streams. Lacking the majestic form of a Douglas-fir or a furrowed Black cottonwood, the inelegant willows are nevertheless a beneficial addition to the forest. In early spring, their pinkish cast adds cheerfulness and promise to the gray landscape, while during the summer season the rich green of their leaves softens the sharp angularities of the glacial streams.

Pacific willow leaves

Appendix B
Reproduction

Gymnosperms

In the animal kingdom, a variety of strategies enhance reproductive success. The males of many bird species advertise their presence to both prospective mates and competing males with colorful plumage, elaborate displays, and vocalizations. In contrast, female birds are often muted and camouflaged. Size disparity between male and female in many mammalian species increases the male's ability to exert dominance, establish territories, and fend off rivals. Some of these adaptations might be regarded as finely tuned to the environment, while others seem extravagant or risky by comparison. The final test is manifested in the reproductive success of the individual and the continuance of the species.

Subalpine fir male cone

Unlike most animals, plants are stationary, a characteristic that not only distinguishes them, but subjects them to unique stresses as well as opportunities. This constraint is such an important reality of the plant kingdom that you might expect a method of propagation different than animals. This is not the case, however, for although asexual reproduction is employed by many plants, like the animals, most plants reproduce sexually, utilizing both male and female roles. Although not a behavioral quality in the animal sense, the basic biology and method is remarkably similar.

Thus, the reproductive cycle in plants includes the maturation of male and female parts with seed production and viability often varying considerably from year to year. In the gymnosperms, it is dependent, along with other parameters, upon the timing of the maturation of male and female cones.

Pollen

Exposed to filtered sunlight and devoid of living things, the barren land stretched from horizon to horizon, an open invitation to a crowded sea.

Although life existed in the seas early in Earth's history, emergence from the waters was slow. Not until approximately 400 million years ago was the transition made, probably by a now extinct algae. Once begun, however, the tentative holds on the land developed into a full scale invasion. To adapt to the demands of dry land, the algae gave rise to the bryophytes and the tracheophytes. It is these two that possess the land today.

The modern range of the bryophytes and tracheophytes reflects a set of traits that evolved as a response to the peculiarities of a land existence. The bryophytes, a group which includes modern mosses and liverworts, retained their ancient dependence on free-standing water for reproduction. This method, which may have further evolved during a rainy era, imposed severe restrictions on the bryophytes. That reliance was limiting in dry or cool conditions, and, ultimately, in most land environments. Today, the requirement of water for reproduction continues to restrict the bryophytes.

Whitebark pine pollen cones

With the ancient bryophytes essentially unchanged, it was left for the tracheophytes to adapt to the new conditions. Since water was less readily available on land, the tracheophytes evolved reproductive strategies that no longer directly depended on this element. The gymnosperms, a tracheophyte subphylum which includes the conifers, dispensed with water transport entirely for fertilization. Wind-borne pollen produced within the male structure became the fertilizing agent for the egg. Providing further protection, the products of the union and a store of nutrients were enclosed within a protective seed coat.

The pollen grain is a complex structure with a range of adaptations. Varying from species to species, the outer surfaces of the grains are geometrically varied, an adaptation that may enhance the ability to stick to a surface. Dispersed by outside forces, contact of this microscopic particle with a female counterpart of the same species is as likely to meet with failure as success. To increase the odds, thousands of pollen grains are produced.

Noble fir pollen cones

The development of the pollen grain resulted in different reproductive strategies. A pollen grain may be produced on the same flower as the female, within its own flower on the same plant, or on a separate plant altogether. This latter method reduces self-pollination, thus increasing genetic diversity. In many species, the same goal is reached by different maturation rates for the pollen grain and the female ovule.

The female part contributes to the fertilization process by producing a sticky fluid which catches the pollen grain, further triggering its development. This includes production of a tube that extends into the ovule where the pollen grain deposits dual cells for a double union with the egg. Successful pollination results in a hard seed, produced within the female cone. Germination and successful development of the seedling now depend on a combination of new physical and biological forces. The pollen's function is complete.

Ovule

Among the oldest of plants, the cosmopolitan algae are prolific today. From the earliest unicellular species, to the later more complex forms, the algae have adapted to a multitude of land and water environments. There are several algae phyla, each with a distinctive coloration, and some are very well known. The brown algae are the seaweeds of cool waters, while the most common red algae populate the tropical seas. The most diverse group, however, are the green algae; only they transitioned from water to land.

Algae employ both asexual and sexual reproductive strategies, and many species utilize a mechanism common to modern plants. This is the alternation of generations, a strategy that employs two distinct plant forms: the sporophyte (asexual), and the gametophyte (sexual) generations. The sporophyte form is the product of a sexual union. It, in turn, produces spores, from which the typically smaller gametophyte develops. The gametophyte organism produces egg and sperm cells. Fertilization takes place within the watery environment on the gametophyte; receiving no further protection or nourishment this union develops into the sporophyte, and the cycle begins again.

Noble fir cones

Size difference in the alternate forms is the rule. The gametophyte is the larger structure in the bryophyte phyla, while in the tracheophytes, which includes most modern plants, the gametophyte is reduced in size. Ferns are older tracheophytes, and although they maintain two distinct forms, the gametophyte is much smaller than the sporophyte. The gymnosperms, also tracheophytes, have carried the evolutionary trend even further, with the tiny gametophyte reduced to total dependence on the sporophyte.

In the conifers, the female gametophyte develops within an ovule, an organ attached to the cone scales, the sporophyte structure. The immature scales are soft and typically green in color; following fertilization and maturation, the female cone acquires a hardened, woody appearance. Within the ovule, cellular division produces four spores, one of which survives as a tiny gametophyte which develops into one or more eggs. The process may take as long as a year, during which time the male counterpart (pollen) maturation proceeds in a manner to coincide with the ripening egg. As this process nears completion, the ovule secretes a sticky substance upon which the pollen is caught and, drying, is drawn into the egg through a tube. Following fertilization, the contents of the ovule may develop into a seed; if this occurs, the ovule enlarges and hardens into a protective seed coat. Within, a tiny embryo (the sporophyte) carries the genes of the parent tree.

Successful fertilization does not necessarily ensure viable seed development. The number of seed-bearing cones and the quantity of viable seeds may vary greatly depending on the particular species, the age of the parent tree, or on a cyclical basis. Environmental factors are important, and an enforced period of dormancy may affect development. The method is obviously highly successful. No longer dependent on water for reproduction, and adapted to a wide range of conditions, the gymnosperms remain widespread today.

Seed

Preceding the angiosperms by more than a hundred million years, gymnosperms first embodied a radical departure from the alternation of generations mechanism. The word "gymnosperm," meaning naked seed, refers to the fact that the product of fertilization develops and receives its protection and nourishment within an ovule. Eggs are produced which, when fertilized, undergo cellular division to form the young plant. During this process the ovule matures and hardens into a protec-

Pacific silver fir seeds

tive seed coat which surrounds the developing young plant. The seed is attached directly to the cone scale with no surrounding fruit or tissue, as is the case with the angiosperms. Drying, the seed-bearing cone relinquishes its seeds, either by disintegrating in place, releasing the entire scale, or by opening sufficiently to disgorge the individual seeds, leaving the cone intact. The seed falls either near the parent tree or, with the aid of a hard wing or two, is dispersed some distance by a breeze. Some gymnosperm seeds are large enough to attract animal carriers, but many are tiny and fall unnoticed.

From the smallest seed (110,000 to make a pound!) comes one of the largest trees, the Western redcedar. Even the colossal Douglas-fir traces its beginnings to a tiny seed. No matter how small, the seed is a complete package, containing not only the embryonic plant and root, but a small amount of nourishment. Many seeds require a period of dormancy before germination, sometimes including an extended cold spell. For most, viability decreases with time, although some may lay dormant for years.

The chances of germination of any seed are small, and although some food is available, success, for the most part, dependent on external factors for sustenance and growth. Derived from the union of pollen and egg, the seedling will face many obstacles to survival.

Angiosperms

Pollen

Clouds of pale color fill the air and dust the tops of cars.

Many tree species produce identical copies of themselves through vegetative propagation from roots or branches. There are several advantages to this method. The precise timing required for sexual reproduction is less critical with cloning, and nutrients are more readily available within the environment of the parent tree. The offshoot carries identical genes of the parent, already a proven success. Sources of stress such as seedling diseases and inadequate moisture may be reduced for the vegetative shoots, and a supply of genetic material is always available.

Seedlings and clones

However, the nutritional and protective advantages for a clone are countered by the possible disadvantages in identical genetics. Sexual reproduction increases genetic diversity, incurring possible survival advantages to a large organism, particularly in a changing or stressed environment. In addition, dispersal range is generally reduced with the vegetative method. Competition close to the parent tree may be limiting; adaptation to different conditions, such as shade intolerance, might favor the wider dispersal available to seeds.

Red alder male structure

In fact, most trees are the product of a sexual union.

The male contribution to this union is packed in tiny pollen grains. These mature in a specialized structure – the stamen – which consists of an anther and a filament. As with the gymnosperms, the angiosperm stamen may reside on the same flower as the female part, on its own flower, or on a separate plant. To reduce self-fertilization, and thus increase diversity, female and male parts often mature at different rates. In this case, the pollen must disperse to a neighboring plant with mature female parts, structures often located a considerable distance away. Reducing the risk of self-fertilization in this manner places an added importance on the means of dispersal.

In addition to utilizing physical forces for dispersal, the more recently evolved angiosperms have another innovative means at their disposal. This is the angiosperm flower, an eloquent structure designed and refined for a complex partnership. Those attributes of flowers that are most appealing to humans--the distinct colors, smells, and forms--attract mobile animals. Specialized insects and birds are drawn to the flower's center where a sweet and nutritious nectar may be proffered. The expenditure of energy for flower and nectar production is costly for the plant, and thus is not given without reciprocity, and so the laden insect departs dusted with pollen grains. Arriving at another flower, the insect inadvertently brushes off some pollen, depositing the grains on the receiving female stigma. Even when nectar is not proffered by the flower, the pollen may be attractive in its own right.

In the microscopic world of the pollen grain, functionality is not dependent on beauty, but it may be dependent on form. The outer surface of the grain is typically roughened with an angular geometry. This form is often related to the physics of transport, aiding in the journey to the female organ, either by wind or by animal transport. After contact, the pollen grain expands into a tube which extends downward into the ovary where the male cells are deposited. Fertilization follows, and with it the beginnings of a seed.

Ovule

Packed within the center of the angiosperm flower are one or more specially modified leaves, the carpels. The base of each carpel is swollen into an ovary, the female reproductive organ, and attached to this structure are one or more ovules. It is within the ovule that fertilization and subsequent development takes place. The ovule consists of a wall (integument), later to become a hard seed coat, and a cell which divides into four parts. This division occurs while the flower is in its budding stage. As the flower matures, one part reaches maturity at the expense of the other three. This fourth piece now divides again, producing seven cells. Only two of these cells are utilized in the completion of the reproductive cycle; one egg (with its single nucleus), and a cell containing two nuclei. Both cells are fertilized by the two sperm packed in the pollen grain, with the extra nuclei forming a three-part union. No other group of plants employs such an exotic mechanism.

From the small seed .....

Although the reason for the three-part union is not understood, the product is not wasted, as it matures into a nourishing food (the endosperm) for the fertilized egg. Except for the cell division, little structural change takes place in the ovule until fertilization. After this event, the organ serves as the developing site for the young plant. Multiple cell divisions create an indistinct juvenile plant, complete with a tiny root. With its mission nearly complete, the ovule sacrifices itself to the new shoot, slowly drying and hardening its outer covering into a protective seed coat. The newly formed seed, complete with its tiny enclosed plant, waits for transport.

Seed

The warm times were long, and the monsters grew larger. But the restless earth moved, and a cooler time began. It was an ending and a beginning. Rele-

gated to a minor role in the past but now survivors in a world more suited to their strengths, tiny mammals and primitive plants poised to take possession upon the giants' demise.

As the age of dinosaurs came to an end, an accelerated evolution began in both the plant and animal worlds. This new era in Earth's history, the Cenozoic, was characterized by much cooler and drier conditions than the preceding Cretaceous, a change that was to favor both mammals and the primitive angiosperms. Responding to the new environment, angiosperm evolution accelerated, leading to the modern flowering plants, represented by over 250,000 species today.

Similar to their gymnosperm ancestors, the reproductive process of the angiosperms is independent of water. Likewise, successful fertilization results in development of a seed. Functionally, both gymnosperm and angiosperm seeds carry the prototype of a new plant. But at this point the resemblance ends.

Scouler's willow capsules and seeds

Differing from the gymnosperms, the angiosperm seed is surrounded by the "endosperm," a nutritious tissue with an unusual origin. Borne of the union of two female cells and one male cell, the endosperm is unique in the living world. This structure may provide food to the developing seed, or it may be present in the completed product as a nourishing tissue. In this latter case, the enriched endosperm may be used by the seedling, or be consumed by animals.

Angiosperm seeds differ from gymnosperm seeds in another important way. The seed is often encapsulated within a fruit, a nourishing and appealing tissue for the consumer, and an ingenious device for the plant. A means for dispersing the seeds, which often pass unaltered through the host animal's system, the fruit has freed an-

Bitter cherry fruit

giosperms from dependence on strictly physical forces, and contributed to their success and proliferation.

The angiosperm fruit develops from the multiple-layered ovary. Angiosperm fruits vary from familiar types such as apples, to more exotic plans, like sticky burs which, clinging to animal fur, are carried on the host. Some angiosperms attach wings to the fruit (rather than to the seed); others pack an explosive which, at the proper time, causes the fruit to fling its contents in all directions.

In terms of species, the angiosperms dominate the plant kingdom. However, there are few angiosperm tree species at Mount Rainier, and they are, for the most part, confined to the lowlands. Their presence is nevertheless apparent, from the stark limbs of a cold December morning, to the sunny July afternoon when the air is suffused with light from pale leaves, and the alpine meadows burst with flowers, another unique contribution of angiosperms to the natural world.

Flowers

With little regard for human concerns and visitations, the ephemeral flowers bloom and die within days, answering only to the shortness of the season and the need for renewal.

When dinosaurs dominated the animal world, the warm climate sustained a diverse and exotic plant community. Some of these plants were related to familiar species of today, while others were destined for extinction, leaving remnants only in the fossil record. Lush and densely packed communities of these ancient plants provided the food necessary for the survival of many dinosaur species. But if it were possible to step backwards in time and visit the dinosaur's world, we might sense there was in incompleteness in the plant community. An abundance of cool greens would abound, but the vibrant reds, oranges, and yellows would be missing. There would be no flowers, for these are the invention of the angiosperms, a group that was barely represented during this time. Destined to dominate the plant world, the angiosperms proliferated after the demise of the dinosaurs.

Bitter cherry flower

A prolific collection of exotic names describe the configuration of individual and collections of angiosperm flowers. Groups become "racemes," "clusters," "inflorescences," "corymbs," "peduncles," or "panicles"--blooms are "florets," "cleistogamous," "complete," or "imperfect." The configuration of the deciduous tree flowers at Mount Rainier are described by these definitions; for example, Bitter cherry brightens the forest with five-petaled white flowers arranged in a corymb, or flat-topped cluster. Bigleaf maples display multiple yellow blooms in a raceme, a single stalk of linear flowers stacked on top of each other, the blooms attached by short stems.

Compared with the bright flowering herbs, tree flowers may seem insignificant. However, on closer examination these less spectacular flowers reveal a delicate beauty. The "sepals" (outer petals) of Vine maple flowers are purplish; the tiny, yellow-green Bigleaf maple flowers are borne in luxuriant clusters, and the dark Bitter cherry offsets its white clusters with protruding yellow stamens. Brightening the streamside forest with a pink glow, thousands of Red alder catkins hang petal-less, ready for the ripening woody female counterparts. Overhanging Black cottonwoods bear grapelike clusters of purplish flowers; later in the season these blooms will mature into the tufted seeds that give the species its name.

Bigleaf maple raceme

The inconspicuous willows have traded conventional petals for a whole new scheme. Just beyond the Nisqually Glacier overlook on the road to Paradise, the rocky slope is populated by masses of Slide alder limbs and the thinner branches of willows. The thin substrate apparently does not deter the alder or willow, although the latter is primarily confined to the wettest areas. Walking this road in April, the visitor is treated to a willow display of fuzzy blooms borne on stark twigs that extend tendril-like from the slope, all in close proximity to a temporary waterfall. Willows have forfeited the petal idea entirely with male and female catkins that resemble each other except for the long stamens which protrude from the former.

The nectar and pollen proffered by the angiosperm flowers attracts honeybees and other insects. Enlivening the quiet forest with their humming, the active insects confirm that the angiosperm flower is well-represented in the trees of Mount Rainier. Functionally equivalent to and equally as important as their showier counterparts, the deciduous trees offer the first hint of spring, and a prelude to the promise that summer will come after all.

Appendix C
Visiting The Forests

Overview

Two major highways provide access to the subalpine region at Mount Rainier National Park. On the northeast side of the mountain, the road to Sunrise follows the White River from the lowlands to an elevation of over 6,000 feet. The road traverses a dramatic landscape of deep canyons, dark cliffs with a fresh appearance, and forests which range from old growth to diminutive subalpine shrubs. This is the so-called "dry" side of the mountain, where timberline is higher than on the south side.

The road to Paradise also traverses a regime of contrasts on Mount Rainier's south flank. Thick lowland forests yield to sites with major disturbances as recently as a half-century ago; a glacial valley that reflects the glacier's recent retreat is crossed by a high bridge, and the beautiful and ancient Tatoosh Range provides a stunning backdrop. Paradise is lower in elevation than Sunrise by nearly a thousand feet, but the mountain is closer here, and the visitor can hike to the edge of active glaciers. Some of the tree species found on the eastern slopes are absent at Paradise, perhaps in part due to the high snowfall of this region.

This appendix provides a detailed description of the forests that border the highways to Sunrise and Paradise. Specific stops along each road, with mileage to the stop from the Park entrances, are discussed, with an emphasis on the species present. The intent is to provide more specific information on the species and their communities. In this way, whether the visitor pauses at one stop or more, they are provided with the opportunity to meet first hand the forests of Mount Rainier.

The Highways

Road to Paradise – beginning at the Nisqually Entrance

Stop #1
Mileage: 0.0 miles
Elevation: 2,000 feet
Nisqually entrance

Immediately after crossing the Park boundary on the south side of the mountain, a dense lowland forest dominated by Western redcedar, Western hemlock, and Douglas-fir crowds the road. The forest is particularly dramatic for the contrast it provides to the younger second growth forest outside the Park. A parking space on the right hand side near the entrance provides a good stopping place. Here there is opportunity to stroll through a mixed Western redcedar and Douglas-fir community. The ground forms a little hollow, undulating gently between the road and the Nisqually River. Old stumps attest to a selective logging operation; the remaining Western redcedar range between two and three feet in diameter. They dominate this wet forest, sometimes forming nearly pure stands. Also sharing the space are fewer numbers of Douglas-fir, including one over four feet in diameter. The undergrowth is sparse where closely packed trees form a dense canopy. Along an old road parallel to the highway, a few Grand fir saplings thrive in the more open areas, but in the denser locations there is little evidence of new seedlings.

Western redcedar is found not only on the lowest ground, but throughout this strip of land. This species is present from the Nisqually entrance, at an elevation of 2,000 feet, to approximately 3,000 feet. It is most abundant in the first three or four miles, where swampy areas on the right hand side of the road provide excellent habitat. The drooping pale green foliage and red shaggy bark make identification of this species easy.

Stop #2: *"Changing Composition"*
Mileage: .8 miles
Pull-out, Left Hand Side
Elevation: 2,000 feet
Forest: Lowland-- Douglas-fir, Western redcedar

Leaving Sunshine Point, an old forest dominated by Douglas-fir and Western hemlock crowds the road and populates the hills. Douglas-fir ranging from two to over five feet in diameter, and a Western hemlock community of multiple ages define this typical lowland forest. The result of abundant rainfall is evident on the uphill slope and across the road where a swale of Vine maple and Red alder remains wet throughout the year.

Recognizing Douglas-fir is made easier by the many cones scattered about on the ground. The three to four inch tan cones are distinctive with their so-called "mouse tail" bracts. The old firs are covered with the characteristic reddish-tan bark, deeply and widely grooved.

Stop #3: *"Lowland Swale"*
Mileage*:* 1.43 miles
Pull-out, Left Hand Side
Elevation: 2,040 feet
Forest: Lowland— Douglas-fir, Western redcedar

Upstream from this stop, Tahoma Creek runs rampant; ripping up forest and road alike, the stream rushes beneath the bridge to its confluence with the Nisqually River a short distance away. The creek passes through a wet lowland forest of Western redcedar and Western hemlock, accessible by path. Western hemlock is easy to identify with its lacy foliage and one-inch cones attached to the branch tips and scattered about the forest floor. The canopy is more closed here than at Sunshine Point, with abundant woody debris in this less recently disturbed site. Pacific silver fir is a minor component at this elevation; most of the seedlings are Western hemlock, the climax species.

Stop #4: *"Old Douglas-fir"*
Mileage: 1.70 miles
Pull-out, Left Hand Side
Elevation: 2,180 feet
Forest: Lowland- Douglas-fir, Western hemlock

Vine maple brightens this somber forest of straight Douglas-fir and Western hemlock. Lit by a southern sun on sloped terrain, the forest reaches upward above an understory of Oregon grape, moss, fern, and bright yellow violets, as well as a young community of Western hemlock. Look through this young forest to see a big Douglas-fir, nearly 6 feet in diameter. Like many old firs, the thick bark of this giant is scarred with the black marks of fiery battles. The age of this fire may be approximated from the lack of evidence of burn on a smaller, (three foot diameter) Douglas-fir.

There is considerable downed timber in this forest. Western hemlock and Vine maple are common; beneath grows an understory of red huckleberry, and red elderberry. Salal, often an indicator of impoverished or dry soils, also flourishes here.

Perhaps most surprising of all is the presence of a young Grand fir near the clearing. This species typically grows along sandy stream banks (see Sunshine Point description); its presence may indicate that an old Grand fir grows nearby.

Stop #5: *"A Mud-Altered Terrain"*
Mileage: 3.4 miles
Parking Lot, Right Hand Side
Elevation: 2,340 feet
Forest: Lowland—Disturbance

In 1947 a mud flow spawned on the flanks of Mount Rainier swept down the Kautz River Creek valley. Scouring and widening its bed, the river of mud altered and sterilized the terrain, and uprooted whole trees, depositing the debris along its destructive path. The lush forest that lined the banks of the river became a memory, replaced by a soupy mass. A stand of large dying trees provides a reminder that even size was no defense against suffocation. In its wake, a new and varied forest took form.

A large parking lot at the site is bordered by young Douglas-fir and

Western hemlock, and Western white pine grows in the median near the road. Near the trailhead, Black cottonwood and willow borders the trail that leads to an explanatory sign and an overlook of the valley. At this location, the path of the flow can be determined as leading from the right; the south side of the stream (left hand side) is bordered by the old forest. An excellent trail traverses the valley, and provides an opportunity to study the reforestation of a landscape completely stripped by a major disturbance.

To some degree, a mudflow resembles other major disruptive forces, such as fire. Fire, however, has varying degrees of completeness. The destruction from the mudflow was extensive; any pioneering trees must deal with the resultant nutrient loss, rocky substrates, and hardpan. A potpourri of species, including Douglas-fir, Western white pine, Western hemlock, Grand fir, Red alder, Noble fir, and even a few Pacific silver firs are have invaded the raw land. Some appear impoverished, growing slowly with shortened needles, and heavy coats of lichen.

The widened river valley is under the influence of the mud alterations as far downstream as its confluence with the Nisqually River, a half-mile away. Approximately nine miles upriver, the creek is sourced at the base of an ancient glacier of the same name which stretches to near the mountain's summit, over 12,000 feet above.

The approximate edge of the mud flow is easy to determine from the many snags. Looking up the road, it is evident that a path at least a quarter of a mile wide was cut here. Near the parking lot, snags are ghostly reminders of the old forest behind the younger community at the edge. Their size indicates that the former was old, and probably consisting of Douglas-fir and Western hemlock. The depth of the flow and its path can be estimated from the cut below the bridge. Snags within the current stream bed indicate that a new bed was created.

Stop #6: -"*Fallen Trees Above the Kautz Creek Mudflow*"
Mileage: 3.6 Miles
Pull-out, Left Hand Side
Elevation: 2,400 feet
Forest: Lowland

Back into an old forest spared from the mud flow at Kautz Creek, this area is populated with Douglas-fir and Western hemlock. The fir are old trees, with the largest ranging from 6 to over 7 feet in diameter. Western hemlock vary from 4 to over 5 feet in diameter, indicating the long term presence of this species.

The younger trees are Western hemlock and Western redcedar interspersed with a few Pacific silver fir. At this elevation the climax forest tends to be dominated by Western hemlock, and here an abundance of nurse logs provides this species with good germination opportunities. Evidence of logging is present in several stumps. This is an aging forest, dominated by Western hemlock interspersed with a few Pacific silver fir. Several moisture-loving understory species, including Devil's club with its large leaves and forbidding thorns, indicate wetland conditions in a region of abundant rainfall.

Stop #7: *"Wetland Pacific silver fir"*
Mileage: 3.85 miles
Pull-out, Right Hand Side
Elevation: 2,440 feet
Forest: Lowland

Similar to the previous stop, this location provides access to an ancient lowland forest. A small stream undulates through flat land that may at times be inundated. This slight depression is bordered with one foot diameter Pacific silver fir. Large Douglas-fir, including some over 7 feet in diameter, and Western hemlock over two-and-a-half-feet in diameter populate the drier areas. On the left hand side of the road an ancient Douglas-fir, undoubtedly one of the largest in the Northwest, boasts a greater than 10 foot diameter.

Stop #8: *"Twin Firs Trail — Magnificent old Dougs"*
Mileage: 3.9 miles
Pull-out, Left Hand Side
Elevation: 2,460 feet
Forest: Lowland

Moss-covered logs, eight foot diameter Douglas-firs, stout Western hemlock, the songs of varied thrush-this is an old forest. A short trail traverses this beautiful forest. However, it is not dominated by ancient trees alone. New opportunity is provided in downed trees and in the holes left by uprooted giants.

Small blooms of twinflower, vanilla leaf, and rattlesnake plantain dot a luxuriant understory, and several species of ferns, including deer fern, sword fern, oak fern, and bracken fern contribute to a lush environment. Devil's club grows in wet seeps, and shrubs are common.

Many other tree species populate the Twin Firs forest. It is a complex community, home to many lowland species, including Pacific yew, Western hemlock, Douglas-fir, Western redcedar, Pacific silver fir, and Vine maple.

Stop #9
Mileage: 4.3 miles
Pull-out, Right Hand Side
Elevation: 2,580 feet

The Nisqually River shines through a forest of tall and slender

Western hemlock, with a smaller number of Western redcedar. Across the road, a stand of one to two foot diameter Douglas-fir occupies a rise; if you look up, you can see the light brown cones hanging in great numbers from the tips of the branches. A few Western redcedar grow here, although they seem more common along the parking strip.

Young Douglas-fir grow on the right side of the road also, and the entire forest appears to be more densely packed than the one at the previous stop. There are many fallen trees here with few saplings in the understory.

Close to the river, on old road traverses the bank, creating an opening. A lush herbaceous layer that includes red huckleberry, pippsissewa, wild blackberry, and other species forms an understory beneath small trees dominated by Western hemlock, with an occasional Pacific silver fir. Downhill a few Western redcedar grow.

This forest is a classic young lowland forest, composed primarily of tall Douglas-fir and Western hemlock. Noble fir appears to be absent, although this species does occur at this elevation (see previous stop).

Stop #10
Mileage: 4.6 miles
Pull-out, Right Hand Side
Elevation: 2,620 feet

Although a forest may appear to be constant in composition, transitions are often quick. Here is found a community of Douglas-fir much older than that at the previous stop. Looking back down the road and across, an aged Douglas-fir occupies the slight slope. At the end of the parking lot another six foot diameter individual grows. Along a trail to the right, a foot-and-a-half diameter Western hemlock reveals its origins on an old nurse log which still protrudes between the hemlock's roots. There are many large downed trees here.

This forest is easily traversed, except perhaps for the wet swale on the right, which reveals itself with skunk cabbage, Devil's club, and Red alder, species indicative of wetland conditions. Following the trail towards the river, a hollowed out Western redcedar is still alive. Beneath it lies a bed of huckleberry and Oregon grape. Uphill from the cedar a Western hemlock reveals mistletoe damage in its stunted lower branches. Old nurse logs near the river support four foot tall Western hemlock saplings.

Stop #11
Mileage: 4.75 miles
Pull-out, Right Hand Side
Elevation: 2,650 feet

A good view of the Nisqually River as well as access to a forest that is a mixture of the common lowland species and a small number of Pacific silver fir makes this an interesting stop. Western hemlock is the dominant climax species here, with many one to two foot diameter trees growing along the valley walls. At the entrance to the parking strip several sizable Red alders add a bright contrast with their oval leaves and whitish trunks. A large Western redcedar grows near the upper end of the parking lot.

The valley bottom is sprinkled with uprooted trees, silent evidence of the turbulent nature of the river. The broad, limbed tops of several large Douglas-firs provide a dark backdrop upstream and across the river. An old trail leading to the river is bordered by young Pacific silver fir, while an old Douglas-fir completes the scene. On the other side of the highway, a dense forest composed primarily of multiple-aged Western hemlock populates the sloped terrain, and a wet depression supports a stand of 30 foot tall Pacific silver fir. Western hemlock dominates the forest here, but for a reminder at how quickly forest composition can change, look back down the road to the big Douglas-firs.

Stop #12
Mileage: 5.6 miles
Pull-out, Right Hand Side
Elevation: 2,680 feet

Along the road Western redcedar populates the open area against a backdrop of tall Western hemlock, while the Nisqually River flows between a border of Red alder. Across the road a border of ten foot tall Western hemlock saplings and a Douglas-fir grow in an apparently recently disturbed area. Douglas-fir three to four feet in diameter also populate the forest. Most of the young Douglas-fir here are confined to the rocks and sand of the river bed, a substrate undoubtedly nutrient poor.

In the middle of the river bed, dying Black cottonwood may indicate a change in the river's course. Downstream, tall Western hemlock and Western redcedar hint at a more aged and stable landscape. A side channel is the site of a robust Red alder community, but Grand fir, a species previously noted in this type of habitat, is absent. Above this flat area, many one foot diameter Western hemlock dominate the bank, and, to complete the scene, a backdrop of tall Douglas-fir populate a middle section of the old river bed.

Stop #13
Mileage: 5.8 miles
Pull-out, Right Hand Side
Elevation: 2,700 feet

At the lower end of the parking lot two old Douglas-fir grow on elevated terrain, while below the road, a wetland habitat supports a community of Red alder amidst on undergrowth of skunk cabbage and Devil's club. Across the road, Western hemlock two to three feet in diameter form a stand sufficiently dense to block available sunlight.

Young Western hemlock and Pacific silver fir grow amongst many fallen trees. In general, the forest gives the impression of greater age than that at the previous stop; it is fairly open with large downed trees, varying age Western hemlock, and some large Douglas-fir.

A small stream flows through an open area, but in spite of the appearance of good habitat there is little undergrowth along the banks. Possibly sunlight is too restricted by the dark canopy, or perhaps a slide removed a thriving plant community. Six foot diameter Douglas-fir scarred by fire give an indication of the history of this forest. On top of a nurse log a new generation of Western hemlock thrives.

Stop #14: "*Survival In A Hot Springs*"
Mileage: 5.9 miles
Elevation: 2,800 feet
Longmire

The big trees have disappeared, and are replaced by younger individuals which give an impression of openness, even though they are closely packed. In front of the lodge two species of pine, Lodgepole and Western white, grow on ground disturbed by construction. This forest is a meeting ground of lowland and middle elevation species, and many conifers are present, including Douglas-fir, Western redcedar, Western hemlock, and Pacific silver fir. A mixture of Red alder, Black cottonwood, and Slide alder populates the river banks.

Across the road from the buildings at Longmire, bubbling hot springs at several locations in a wide meadow are a reminder of the volcanic nature of this region. Lodgepole pine grows in the more elevated regions of the meadow, displaying an adaptation to the chemically altered substrate. Lodgepole pine not only makes an invasive attempt at the meadow, but grows along the gently sloping terrain between the meadow and the road. To the north, the meadow is bordered by a dense forest dominated by Western hemlock and Western redcedar.

A Grand fir sapling ekes out an existence near the Longmire ranger station; it is doubtful that this species is found above this location. The sandy and open river banks are no longer present along the Nisqually, and thus the preferred habitat is not available. Not many miles from Longmire, the Nisqually River pours from the snout of the Nisqually Glacier. In that region, pioneer species struggle to establish themselves on the sterile glacial till.

Note: Average tree size increases again as the road curves around the base of Rampart Ridge above Longmire. This ridge stretches like a wide backbone in a northeasterly direction towards the mountain. The dense forest hides the view of the overhanging ridge, but its influence is apparent in the depth and persistence of the snowpack. At about a mile-and-a-half past Longmire the road has climbed to approximately 3,000 feet, and at this transition between lowland and middle elevation forest, a new species grows. This small tree is Alaska cedar, an inhabitant of wet and rocky sites throughout the middle altitude and subalpine forests. It gives itself away with its shaggy, loosely hanging gray bark and droopy foliage. In this region, starting just below the Nisqually Glacier overlook, another transition occurs between Red alder and Slide alder. The two commingle almost to this point, but Slide alder grows readily in steep terrain, while Red alder prefers wetland or streamside habitats. The other common lowland deciduous species disappear with the increase in elevation.

Stop #15: *"A Pioneering Forest"*
Mileage: 8.2 miles
Left hand side
Elevation: 3,200 feet
Cougar Rock

The forest at Cougar Rock is similar to the forest at Kautz Creek in the sense that it was established on terrain devoid of life. In both cases, the condition of the landscape is attributed directly to the presence of Mount Rainier, but the forest here is much older, dating from the recession of the Nisqually Glacier. Western hemlock, Western redcedar, and Western white pine are present, but Pacific silver fir dominates here in the heart of the middle elevation forest, replacing the old, pioneering Douglas-fir.

With many of the characteristics of an old growth habitat, the forest at Cougar Rock nevertheless gives the impression of youth. The Douglas-fir bear the thick bark of old trees, but their size is perhaps smaller than expected, and the maturing Pacific silver fir seem diminished in size. These trees may reflect the infertility of the raw glacial landscape; if this is the case, it would be expected that as the soil is enriched, the saplings will achieve greater size.

Stop #16
Mileage: 10.5 miles
Pull-out, Left Hand Side
Elevation: 3,620 feet
Comet Falls trailhead

Across the road, Noble fir more than a foot in diameter are easily recognized by the needles, which appear short and stunted. These fifty foot tall trees still retain their lower branches. The forest that borders the parking lot is composed of big Pacific silver fir, but the tallest tree (near the upper end of the parking lot), distinguished by the platy bark, is Noble fir. Next to this tree a lone Douglas-fir grows, while Western hemlock is still present at an altitude well above its lowland range. Pacific yew, Western white pine, and Slide alder share the habitat in small numbers; however, the most common understory species is Pacific silver fir, the climax species of the middle altitude forest.

Stop #17: *"A Noble Forest"*
Mileage: 11.6 Miles
Pull-out, Right Hand Side
Elevation: 4,000 feet
Nisqually Glacier Overlook

As the most shade intolerant of the true firs, Noble fir is a pioneer species in disturbed habitats. The largest true fir, Noble fir grows at middle to high elevations, sometimes forming pure stands. Although this species does not grow north of Stevens Pass, it is quite common at Mount Rainier. Young Noble firs resemble Pacific silver fir, while older trees carry their lowest branches high above the forest floor, complicating identification. However, the distinctive, platy, maroon tinted bark provides an excellent field mark.

Looking to the left, a steep buttress may be seen rising to the north; this is an offshoot of Rampart Ridge which apparently escaped heavy glacial carving and protrudes boldly, forcing a bend in the road. The bluish cast of the hillside is not a visual trick; in this case it indicates the presence of a mixed Noble fir and Pacific silver fir forest, one which had its beginnings in a forest fire. Noble firs are moisture-loving species which, nevertheless, prefer well drained soils and are intolerant of excessive cold. They are usually found in the middle elevation zone; as the most shade intolerant of the firs they are confined to newly opened habitats. Thus, the south-facing slope may be an ideal location for a combination of warmth and sufficient altitude to ensure adequate moisture throughout the year. Crossing the bridge, a beautiful noble may be seen on the south side; during the autumn, the barrel-shaped cones provide an unmistakable field mark on the top of the tree. Other species are present here, including a few Alaska cedar, and the site is also noteworthy for the proximity of both western and Mountain hemlock. These are also seen from a birds-eye view, where they grow below the bridge on the northeast side. South and west of the bridge, Ricksecker Point rises nearly a thousand feet.

The bridge over the Nisqually Glacier provides an interesting look up the bowl-shaped valley. Historically, the glacier terminus reached below the bridge, but over the past century, it has retreated far enough so that the ice of the active glacier is no longer in view at this location. The recession has provided opportunity for the pioneering Slide alder along the valley walls, while conifers advance slowly from a dark forested border. This is an excellent location for viewing several middle and subalpine forest tree species, as well as a few stragglers from the lower elevations growing at the upper limits of their range.

Stop #18: "*A potpourri of species*"
Mileage: 12.9 miles
Road, Right hand side
Elevation: 4,240 feet
Ricksecker Point

Recent disturbance along the Ricksecker Point road, and its location at the lower part of the middle altitude zone has provided opportunity for a number of species, including Western white pine, Mountain hemlock, Western hemlock, Douglas-fir, and Alaska cedar.

Fresh slides along the road attest to the impact of storms on the topography. On the left hand side and around the bend from the unstable slope, seven tree species grow in very close proximity. Within a 30 foot diameter arena, Noble fir, Alaska cedar, Pacific silver fir, Western hemlock, Douglas-fir, Lodgepole pine, and Western white pine grow. The presence of so many species is undoubtedly due to the recent disturbance, and it would be interesting to return at a future time to observe changes in composition.

Note: Above Ricksecker

A panorama of the mountain and the steep ridge above the Nisqually Glacier is matched by the grandeur of the saw-toothed Tatoosh Range to the south. Noble fir and Pacific silver fir grow in close proximity, providing an opportunity to distinguish these species. Noble fir needles appear shorter from below, and form a more antler-like configuration than the densely packed silver fir. The needles of Noble fir appear blunt, and rigid, growing tightly against the central twig. Pacific silver fir needles also cover the branch, but they are looser and more spread out in appearance. Small Alaska cedar are easy to identify here with their gray shredded foliage and hanging branches. Across the valley, Slide alder pioneers steep avalanche slopes. The last big Douglas-fir occurs at about 4,000 feet, while above this location the forest is composed of trees no more than two feet in diameter.

Western white pine grows in abundance along the road. The presence of Pacific silver fir increases, approaching a dominating role, until the transition to a Subalpine fir forest at about 5,100 feet of elevation.

Stop #19: "*Paradise!*"
Mileage: 17.9 miles

Elevation: 5,420 feet
Paradise parking lot

Approximately a mile below Paradise, a transition occurs from contiguous forests dominated by Pacific silver fir to the patchy subalpine forests, where Subalpine fir is the most common species. The two species are most easily distinguished by their height and width. Pacific silver fir is much broader and irregular in outline, while Subalpine fir is compact, with a neatly chiseled appearance that seems fitting with an alpine existence. Here the wind-whipped Subalpine fir is subjected to heavy snows and cold. Yet this species populates ridges to the very upper limits of growth, near the edge of the glaciers.

Subalpine fir and Mountain hemlock are easily distinguished from one another. In the fir, dark green brushed-up fir needles cover the branches, each bearing a single white stripe on top and two on the bottom. The crowns are narrow, and the fir often grows in clumps (tree islands), with the youngest saplings on the periphery. In contrast, multi-lengthed Mountain hemlock needles cover the branches in a luxuriant display, and the overall appearance of the tree is much less symmetrical than the fir.

Tracking the ridge lines, Subalpine fir persists to the summit of Mazama Ridge, at nearly 7,000 feet.

Road to Sunrise - White River Entrance

Stop #1
Mileage: 0.0 miles
Parking Lot, Right Hand Side
Elevation: 3,480 feet
White River Entrance

Although the elevation indicates a middle altitude regime, this site is dominated by lowland species. Young Western hemlocks form a dense understory beneath a canopy of old Douglas-fir and Western redcedar, and small Western white pine populate a hillside directly across from the entrance station. The pines are easy to recognize, as these young trees bear the long, soft needles characterize of older individuals. Deciduous species at this stop include Slide alder, a middle altitude species, and young Black cottonwood.

Many of the conifers are large and decorated with filmy lichen. Near the restroom, a grove of Western redcedar and a single old Douglas-fir are part of a dense community. Close by a large, fallen Douglas-fir displays its thick bark and impressive length.

At the uphill end of the parking lot, a Douglas-fir nearly five feet in diameter towers above the forest. The fir is easy to recognize by its size, the covering of thick, reddish bark, and the bracted three-inch cones which litter the ground.

Pacific silver fir, the most common conifer at Mount Rainier and a middle altitude forest species, also grows in this transitional forest. These trees appear bluish when viewed from beneath; the gray, patchwork bark is evident on a one-and-a-half foot diameter individual. Also found in this area are a small Alaska cedar (most easily distinguished from the Western redcedar by its scaly gray bark), and a young Pacific silver fir.

Across the road, most of the young (and many of the larger) trees are Douglas-fir. The road provides an opening for the continuing presence of this pioneering species. The understory of the more shaded areas, however, is dominated by young Western hemlock, the climax species of the lowland forest.

As you leave the station, you will pass through a young forest of Western hemlock, big Douglas-firs on the hillside, and a few Pacific silver firs. Young Douglas-firs populate the disturbed areas alongside the road.

Stop #2: *"Englemann and noble"*
Mileage: 1.4 miles
Pull-Out, Left Hand Side
Elevation: 3,660 feet
Shaw Creek

The needles of both spruce species and Douglas-fir resemble one another, and the Englemann growing here close immediately to the left near the middle of the parking lot might be easily missed. The needles appear a bit stiffer than Douglas-fir needles, and they are prickly. Another difference is that the Engelmann's light tan twigs are knobby at the point of attachment of the needles.

Noble fir also occupies this site, perhaps taking advantage of the openness that is available at the parking strip. As you start into the parking lot, a Noble fir tree is established on the left. Larger than the Pacific silver fir here, the noble may be distinguished from the silver fir by the needles, which have double stomata lines on the top and bottom, and appear brushed up. They are more narrow than those of the silver and tend to be more hooked at the junction with the twigs. Like the Pacific silver fir, the stomata lines give Noble fir a bluish cast, and young trees of both species have smooth, gray bark.

Both Noble fir and Pacific silver fir are species of the middle altitude and subalpine habitats, although the noble is more restricted in range. Many other species of the lowland and middle altitude forests grow at this site. These include Western hemlock, Douglas-fir, and Western redcedar. Their close association is easily seen by scanning from right to left at the parking lot and observing the sequence of species-Alaska cedar, Slide alder, Engelmann spruce, Western hemlock, and a tall Black cottonwood. Thus, no fewer than five coniferous species grow at an elevation that is an intermediate zone between the lowland and middle altitude to subalpine forest.

There are a few large Douglas-fir here, including two at the upper end of the parking lot. Growing nearby is a large Pacific silver fir as well as smaller firs approximately a foot in diameter. Looking back towards the bridge (from the direction you have come) big Douglas-fir and large Pacific silver fir comprise the tallest trees of the canopy. Below the road on the parking lot side, the cedar and hemlock border a creek which may be dry late in the summer season.

Stop #3: *"Middle elevation Douglas-fir*
Mileage: 2.0 miles
Parking Lot, Right Hand Side
Elevation: 3,700 feet

With their deeply grooved reddish bark, the tall Douglas-fir trees at this stop are easy to identify. Across the road a few of the fir are interspersed with Western hemlock, while young Douglas-fir seem to sprout from the rock face on this side.

Looking back down the road on the right hand side back 50 meters or so, a single large western white thrives. This tree is just under four feet in diameter, and grows in a wetland (although at the time of its germination, this may not have been the case) with Western redcedar, and an understory of Red alder, vanilla leaf, false huckleberry and horsetail. In front of the pine, a tall Western redcedar appears to be in much better health than the pine, and to its left is an old Pacific silver fir. Along the parking strip, Alaska cedar commingles with Slide alder.

This is an interesting location because both cedar species (Alaska cedar and Western redcedar) as well as both alders (Red alder and Slide alder) are established. Small Alaska cedar grows in the rocky area in the foreground. The understory is dominated by Western hemlock, although at this altitude, Pacific silver fir begins to play a larger role.

At the upper end of the parking lot, the large leaning tree is an old Douglas-fir, and across the road another old fir with a broken top, survives.

Stop #4: *"Old growth Western hemlock"*
Mileage: 2.2 miles
Parking Lot, Right hand side
Elevation: 3,700 feet
Owyhi Lakes Trailhead

A service road leading towards the White River from a parking lot on the right hand side passes through a stand of large Douglas-fir and Western hemlock. Approximately 100 feet down this road on the left, a large Noble fir with a very straight trunk is recognizable by the plated, maroon-colored bark. The crown of this tree is small compared to its considerable height. At the end of the road is a large Western hemlock with grooved bark that, at first glance, resembles a Douglas-fir. However, the hemlock tends to have more grooves (less distance between the ridges), is grayer in appearance, and less thickened with age. This stop provides an excellent opportunity to compare the two species side by side.

Many of the Western hemlock are dead on top. And, unlike the smooth trunk of the firs, the Western hemlock trunk bears the stubs of fallen branches. The ground is littered with cones, easy to recognize by their small size (an inch or less).

This location also provides access via the Owyhi Lakes trail to an old growth Western hemlock forest growing robustly on a fairly dry site at the upper lavational limit of its range.

Stop #5: " *More old growth forest*"
Mileage: 2.4 miles
Pull-Out, Right Hand Side
Elevation: 3,720 feet

A Douglas-fir nearly seven feet in diameter growing near the parking lot seems particularly large for this elevation. Two large Pacific silver fir also occupy this site, and, looking back down the road and across, a big Western hemlock with an irregular top may be seen.

The White River is very close here, and a walk down a trail towards the river provides the opportunity to see another big tree, this time an Engelmann spruce. This tree is four-and-a-half feet in diameter with a straight, beautiful trunk. Proceeding towards the river, a tall Western white pine may is recognizable across the river; also present here are Douglas-fir, Western hemlock, and large Western redcedar (one is over 5 feet in diameter). In close proximity to the cedar are several large Pacific silver fir.

Thus, at this altitude the forest is composed of a mixture of lowland and higher elevation species, including Western hemlock, Douglas-fir, Engelmann spruce, Western redcedar, Western white pine, and Pacific silver fir. This mixture of such large low and middle altitude species is rare.

Stop #6: "*Subalpine Intrusion*"
Mileage: 2.9 miles
Pull-Out, Right Hand Side
Elevation: 3760 feet
Fryingpan Creek

Behind the rock wall on the south side of Fryingpan Creek, a bushy Subalpine fir grows at the lower end of its range in the Cascades. The stream's origin in the Fryingpan Glacier above Summerland may explain in part the presence of this species. Cold air draining down the valley may have enabled establishment in a micro-climate close to the stream.

Across the road (the same side as the parking lot for the Summerland trailhead) a large Engelmann spruce may be identified by the many cones borne on the branch tips of the upper third of the tree. This spruce is quite different than the spruce at the stop 2, as it appears bluer in color and bushier.

More Engelmann spruce grow near the trailhead; in fact, the tallest trees at this location are spruce trees. As evidenced by the three to four foot diameter individual on the right hand side about halfway along the parking lot, Western hemlock has not yet dropped out altogether Deciduous species include Vine maple and willow species. Another lowland species, large Douglas-fir are still present.

Stop #7: *"Silvers"*
Mileage: 4.8 miles
Pull-Out, Right Hand Side
Elevation: 4,140 feet

In the heart of the middle altitude forest, this is a habitat dominated by Pacific silver fir. The lowland species grow here, too, and, in fact, the large Douglas-fir confirm that lowland species may do very well at higher altitudes. There are big Pacific silver fir at this stop, while Alaska cedar and Douglas-fir inhabit the sloped terrain above. A small Western white pine grows near a grayish Douglas-fir decorated with filmy lichen.

Looking down slope, a fairly dense forest with many Western hemlock in the understory covers the steep hillside.

Stop #8: *"Lodgepole pine"*
Mileage: 4.9 miles
Pull-Out, Right Hand Side
Elevation: 4,200 feet

Lodgepole pine colonizes the rocky terrain at the upper end of the parking lot. To the right of the pine are several young Noble fir, and seedlings of this species grow in the strip near the pavement. Looking back downhill a Black cottonwood displays its heart-shaped leaves; this tree grows 10-15 feet down slope. Both alders are here, growing alongside each other; also present are several young Black cottonwoods, and bluish-tinged Pacific silver fir.

Stop #9: *"Nobles and silvers"*
Mileage: 5.5 miles
Pull-Out, Right Hand Side
Elevation: 4,340 feet

Near the parking lot are three large Pacific silver fir, while several young Noble fir grow at the upper end. Across the road, small Western white pine populate the open bank while big silver firs seem very much at home. Looking back down the road, this is an excellent stop on a clear day to get a view of Mount Rainier.

Stop #10: "*A Beautiful Noble*"
Mileage: 5.7 miles
Pull-Out, Right Hand Side
Elevation: 4,400 feet

Douglas-fir, some dying or with broken tops, share the habitat with Pacific silver fir and Western hemlock. In front of the silver fir is a dying Western white pine, and Douglas-fir occupies the embankment.

Stop #11: *"A Mixture of Species"*
Mileage: 6.4 miles
Pull-Out, Right Hand Side
Elevation: 4,480 feet

At the upper end of the parking lot Black cottonwood, a species usually confined to lowland river banks, has achieved a diameter of more than a foot. Still present are the common Douglas-fir, Western hemlock, and small Alaska cedar. As you enter the parking lot, a beautiful Noble fir with very bluish colored branch tips, and over a foot in diameter, grows in this open habitat. Large Douglas-fir demonstrate the versatility of this unique species, inhabiting a terrain much higher than a lowland designation make indicate.

Stop #12: *"Lava Columns"*
Mileage: 6.5 miles
Pull-Out, Right Hand Side
Elevation: 4,600 feet

An informational sign describes the distinctly hexagonal lava columns protruding as the end of a flow from the hillside. Once again, a number of species common in the lowlands occupy this middle altitude habitat. Douglas-fir and Western hemlock grow alongside small Pacific silver fir, Alaska cedar, and Western white pine.

Across the road a two foot diameter Douglas-fir is covered with bark grayer than considered typical. At the upper end of the parking lot, six foot tall Pacific silver fir grow tightly packed together, and close by a young Western hemlock and Douglas-fir are easily recognizable. Near the end of the parking lot, a big Douglas-fir grows to the right and behind the Pacific silver firs.

A tall but small diameter Alaska cedar grows down the slope behind the group of small Silvers. Across the road and up slope are several dying trees; these may be Subalpine fir.

Stop #13: *"Big Bend"*
Mileage: 7.0 miles
Pull-Out, Right Hand Side
Elevation: 4,800 feet

This interesting stop is near a stream with origins near Sunrise. The pull-out is around the bend. Again (see Stop 6) the direct route from the stream to the subalpine region may explain the abundance of Engelmann spruce, a common species at Sunrise, at this location. There are many Pacific silver fir, and Subalpine fir is also present.

Douglas-fir continues to be represented with big trees on the uphill slope at this location. Alaska cedar grow in the wet, rocky area near the edge of the depression marking the stream's route, and Slide alder thrives on the embankment here.

Stop #14: *"A Monoculture"*
Mileage: 7.4 miles
Pull-Out, Right Hand Side
Elevation 4,840 feet

We have entered a Pacific silver fir forest of varying size. Lenticels form horizontal strips on the young trees, and bear grass makes its appearance in the understory. A dead Western white pine may have succumbed to disease; the ground is sprinkled with the large cones of this species.

Stop #15: *"One Last Doug, and a Grand"*
Mileage: 8.0 miles
Pull-Out, Right Hand Side
Elevation: 5,100 feet

Here is an ideal location for identifying many species, as they are lined up alongside one another, and outline the horizon with large members. Below grows one last small Douglas-fir, a Grand fir, Engelmann spruce, Western white pine, Subalpine fir, and Pacific silver fir, the last two trees of large size. Looking down slope, there are two huge nobles; in the fall these are most easily recognized by the large cones perched on the upper branches. Close to the Noble fir are found two large Engelmann spruce with irregular tops. Left of these are Pacific silver fir with spikes on cones nearly four inches long. The tallest tree at the upper end of the parking lot is an Engelmann spruce.

Across the road the forest is dominated by Subalpine fir, many still retaining their lower branches and pale, non-grooved bark. At the lower end of the gravel parking a very large Noble fir towers above the Subalpine fir.

Stop #16
Mileage: 11.5
Parking Lot, Left Hand Side
Elevation: 6,100 feet
Sunrise Point

The elevation gain of 1,000 feet from the last stop has been through a transition to a patchy forest, one interspersed with meadows and characterized by "clumping" of trees. Subalpine fir is common, although Pacific silver fir shares the habitat. The trees are reduced in stature by comparison to those at the last stop. This is a very altered forest by comparison.

Across the road Mountain hemlock grows on the uphill side. Looking across the canyon, evidence of an old burn is given by the white prostrate trunks and the lower timberline. Many dying trees are present here.

Stop #17: *"Whitebark"*
Mileage: 13
Parking Lot, Left Hand Side
Elevation: 6,400 feet

On the horizon straight ahead the bushy form of a large Whitebark pine may be seen. This is a high altitude species, but it is easily accessed in the Sunrise vicinity. Short, yellow-green needles borne on multiple branches make identification easy; unlike Western white pine (which has dropped out by this point), this species has small, egg-shaped cones.

We are now in a region of meadows interspersed with tree islands, bunches of trees of varying age, small communities of trees that provide protection against bitter weather for their members.

Stop #18
Mileage: 14.0 miles
Parking Lot
Elevation: 6,400 feet
Sunrise

At Sunrise Park a trail leading towards Burroughs Mountain provides an opportunity to see a large Alaska cedar community. There on a steep, rocky north-facing slope the low, yellow-green forms of Alaska cedar cover the ground. Its presence may be attributed to the steepness, for such terrain is prone to avalanches, and Alaska cedar is the most resistant conifer to this type of inundation. On these slopes, the shrublike cedar may appear to grow from bare rock, requiring little soil for sustenance.

Other species here include Whitebark pine, Mountain hemlock, and Subalpine fir. The pine grows well in the well-drained slope above Sunrise. From this location a trail leads to Third Burroughs mountain, where an Engelmann spruce, only a few feet tall, ekes out an existence at an altitude above 7,000 feet.

Glossary

adventitious - Growing from an unusual position, such as roots growing from a stem.
alluvial - The deposition of a stream, particularly over its flood plain.
angiosperm - The flowering plants, bearing enclosed seeds. They are represented by the deciduous trees in the Cascades.
anther—A sack in which pollen is produced.
aril—A fleshy, thickened seed coat, referring to *Taxus*.
aspect - Angle, here referring to the angle of a slope, or the angle presented by the landscape to the sun.
association - A group of tree species regularly occurring together, usually with one dominant.
bryophyte - A division of plants that lacks water-conducting vessels and utilizes alternation of generations for reproduction.
carpel - The female reproductive flower structure.
catkin - A spike of flowers, unisexual, as in the case of male alder catkins.
chlorophyll - Green plant pigment that is the site of absorption of the Sun's energy.
climax forest - A forest in which the dominant species is also the understory replacement species.
clone - A plant derived by asexual reproduction, making it genetically identical to the parent plant.
colloid - Substances with two distinct phases; in the case of soils, one of very small particles with a total large surface area.
conifer - Trees which are members of the gymnosperm plant division, bearing needle-like or scale-like appendages. Most are evergreen, retaining their needles throughout the year.
deciduous - Trees that shed their leaves seasonally.
dioecious - Male and female parts borne on separate plants.
diploid - Having two sets of chromosomes.
disjunct - Separated by a barrier.
duff - The top layer of material on the forest floor.
ecosystem - A community of plants and animals and the environment with which they interact
ectotrophic - Relationship in which the mycorrhizal fungus obtains nourishment by surrounding the tree roots in a sheath.
endosperm- A covering for the angiosperm seed that often contains food materials
endotrophic -Relationship in which the mycorrhizal fungus obtains nourishment by penetrating the tree roots.
fascicle - A tight bundle or a cluster.
filament—The stalk of an anther.

flower - The reproductive organ of angiosperms, in complete flowers consisting of the female and male parts, and surrounding structures.
gametophyte - Plant form that produces reproductive cells by division, each with a single set of chromosomes; the asexual generation.
genetics - Referring to the ancestral physiological lineage of an individual.
glaciation - The covering of land masses with moving ice sheets
glacier - A mass of ice and snow which accumulates, melts, and flows, often covering extensive regions.
gymnosperm - Seed plants with unprotected ovules carried on cone structures. More ancient than the angiosperms, represented by the coniferous tree species.
integument - The coating of the ovule; develops into a seed covering upon maturation.
intraspecific - Between different species.
krummholz—German for "crooked wood." These are trees growing at timberline that are shrublike even when many decades old, with twisted branches and no leading trunk.
leaf - Organ of the tree that is the site of photosynthesis. Usually consists of a blade, a stalk, and attached to the plant at a node. The species at Mount Rainier drop their leaves annually.
lenticel - A raised spot on the woody part of a tree that permits air to reach the interior surface.
megaspore - A female spore which produces the gametophyte* in angiosperms*.
microsite - A spatially small area that may exhibit unique characteristics or plant associations.
monoecious - Plant with female and male parts borne on the same flower.
moraine - A geological structure composed of deposits from a glacier. Often a long ridge that parallels the glacial path, or a terminal mound at the extension of the glacier.
mycorrhizae - Refers to the beneficial association between a fungus and the roots of a plant.
nectar - Sweet and nutritious plant substance attractive to insects. The insect may become dusted with pollen* as it consumes the nectar, thus providing a means of dispersal for the plant.
needle - Photosynthetic organ of the gymnosperms.
nurse log - A fallen tree that is host to seedlings and other plant forms.
order - A classification of plants or animals.
outwash - The debris from a river or other flow.
ovule – Female organ that develops into a seed. Unprotected in the gymnosperms, covered in the angiosperms.
pedicel - The stalk of a single flower.
phloem – In the Tracheophytes, a transport organ for organic and inor-

ganic materials within vascular plants.

photosynthesis - The chemical process which produces organic substances for the plant by utilizing the sun's radiant energy. Takes place within the green chlorophyll of leaves or needles.

Pleistocene - A period of time beginning approximately 2 million years ago and ending with the latest recession of the glacial ice sheets.

podzolic - A soil forming process consisting of leaching of iron and aluminum compounds, resulting in a acidic composition. Common soil process in coniferous forests.

pollen - The male spores produced within the male cone or the anthers of a flower.

regolith - The unconsolidated weathered material layer that sets on unaltered bedrock

samara – A dry, winged fruit that does not open along the seam.

sapling - A young tree.

seed - The structure developed from a fertilized ovum and containing a food reserve and the embryo of a new plant.

seral - A phase in plant community development leading to a climax stage. In general, consists of plants which in general will be replaced by other species.

spodosol - A soil order composed of minerals and organic material, usually in acidic conditions, usually in humid and cool climates. Common in coniferous forests.

sporophyte - The dominant plant form in the gymnosperms and angiosperms which produces spores. The plant tissues are diploid.

stamen - The male organ of a flower, with a stalk like filament and a pollen-producing anther; usually several stamens are present in a flower.

stigma -- The expanded tip of a pistil, which receives the pollen; often on a short stalk called a style.

stipule – An appendage at the base of the leaf petiole. Sometimes surrounds the twig.

stomate -- Pore or opening on a leaf surface, for gas exchange. Plural: stomata.

style -- The slender stalk that elevates the stigma above the ovary in the pistil of a flower.

subalpine -- An elevational zone just below timberline but above the reach of more or less continuous tree or shrub cover.

symbiosis – A mutually beneficial relationship between two organisms.

tap root – Large central root of a tree.

taxol – Chemical substance in Pacific yew bark used in the treatment of cancer.

till – Sediment laid down by glacial ice.

Tracheophyta – A division of plants with vascular tissues through which water and nutrients are transported.

vascular (plant) - Tracheophytes, with vessels for the transport of water and nutrients.

vegetative propagation - Asexual reproduction carried out by plant parts, such as roots or stems.

xylem - In the Tracheophytes, the vessels through which water is transported.

Made in United States
Troutdale, OR
11/03/2026

56844503R00100